"The AI Revolution *is more than just a book— .. _ . y* for anyone navigating the waves of intelligent transformation. With clarity and urgency, the authors spotlight the real-world power of AI across industries, while grounding the discussion in humanity, ethics, and vision. As an educator, technologist, and creative, I'm inspired by this collective of brilliant minds who remind us that thriving in the AI age starts with understanding its potential and embracing its responsibility. This is a must-read for those building, leading, and learning in our AI-powered future."

—MALCOM DEVOE, PhD, Professor of Mathematics & Computer Science; Founder, Devoe Digital Learning

"This book reads like a journey through civilization's next frontier. The authors don't just explore AI's technical capabilities, they reveal its human impact. Each chapter stands alone as a deep dive into an industry, yet together they form a cohesive map of how AI is reshaping our world. It is a brilliant and timely work."

—ANASTASIA POLLOCK, Professor of AI and Emerging Technologies

"The AI Revolution *captures the essence of what makes this moment in history so extraordinary. The book is packed with insights from experts at the frontier of AI development and application. It's both accessible and deeply thought-provoking. It is a must-read for entrepreneurs, professionals, and curious minds eager to be part of the AI Revolution.*

—CASEY LINDBERG, SpaceX, Quality Control Manager

"The AI Revolution *is the spark we need in this pivotal moment of technological advancement. The authors in this book illuminate how AI is already transforming industries and lives, and the author's authentic stories create an emerging universe, unveiling a reality in which technology serves humans. This isn't a book about fearing the future; it's a field guide for thriving within it. This is essential reading for anyone who wants to lead, not lag, in the age of intelligent machines.*"

—KATERINA BOURDOUKOU, Professional Coach & AI Consultant

"*This is not just another book on artificial intelligence. It's a passport into the future. The authors present a global perspective grounded in real-world expertise. The AI Revolution *gives readers an informed and balanced introduction to what's happening in AI across sectors, and what it means for our future.*"

—JOHN HANSON, AI Researcher and Policy Strategist

"*Optimistic, intelligent, and wide-ranging,* The AI Revolution *breaks the mold of doom-filled AI discourse. The contributors offer practical insights rooted in experience across continents and industries. It's impossible to read this book without becoming more curious, and more equipped, to engage with the most consequential technological shift of our time.*"

—SARAH LIN, Engineer, Robotics Coach

THE AI
REVOLUTION

THE AI REVOLUTION

Thriving Within Civilization's Next Big Disruption

Authored by:
Erik Seversen, Jonathan R. Abon, Anupam Agarwal, MD, MPH, Tey Bannerman, Arnaud Blandin, Sean Musch, Michael Borrelli, Jaspreet Chager, Victor Collins, Lisa Goodhand, Basil Hartzoulakis, PhD, James L. Hutson, PhD, Jim Iyoob, Rafał Janczyk, Susmitha Jella, Kent B. Landrum, Georg Langlotz, Glenn Loomis, MD, Alan Macdonald, Tom Mawhinney, Michael Mey, Damien Montessuit, Peggy O'Flaherty, Ravi Pasula, Pramod M. Patke, Carrie Purcell, Philipp Ramjoué, Daria Rudnik, Mark C. Somerville, Georgii Speakman, Sakina Syed, Nicolas Tome, Michael Tutek, Andrew Vasko, Polina R. Ware, PhD, Svetlana Zavelskaya

THIN LEAF PRESS | LOS ANGELES

The AI Revolution: Thriving Within Civilization's Next Big Disruption individual chapters. Copyright © 2025 by Jonathan R. Abon, Anupam Agarwal, MD, MPH, Tey Bannerman, Arnaud Blandin, Sean Musch, Michael Borrelli, Jaspreet Chager, Victor Collins, Lisa Goodhand, Basil Hartzoulakis, PhD, James L. Hutson, PhD, Jim Iyoob, Rafał Janczyk, Susmitha Jella, Kent B. Landrum, Georg Langlotz, Glenn Loomis, MD, Alan Macdonald, Tom Mawhinney, Michael Mey, Damien Montessuit, Peggy O'Flaherty, Ravi Pasula, Pramod M Patke, Carrie Purcell, Philipp Ramjoué, Daria Rudnik, Mark C. Somerville, Georgii Speakman, Sakina Syed, Nicolas Tome, Michael Tutek, Andrew Vasko, Polina R. Ware, PhD, Svetlana Zavelskaya

Disclaimer—The advice, guidelines, and all suggested material in this book is given in the spirit of information with no claims to any particular guaranteed outcomes. This book does not replace professional consultation. Anyone deciding to add physical or mental exercises to their life should reach out to a licensed medical doctor, therapist, or consultant before following any of the advice in this book; anyone making any financial, business, or lifestyle decisions should consult a licensed professional before following any of the advice in this book. The authors, publisher, editors, and organizers do not assume and hereby disclaim any liability to any party for any loss, damage, or disruption caused by anything written in this book.

Library of Congress Cataloging-in-Publication Data

Names: Seversen, Erik, Author, et al.

Title: *The AI Revolution: Thriving Within Civilization's Next Big Disruption*

LCCN: 2025908165

ISBN 978-1-953183-81-1 (hardcover) | 978-1-953183-80-4 (paperback)
ISBN 978-1-953183-79-8 (eBook) | 978-1-953183-82-8 (audiobook)

Artificial Intelligence, Science & Technology, Business, Professional Development

Cover Design: 100 Covers
Interior Design: Dindo Sanguenza
Editor: Nancy Pile
Thin Leaf Press
Los Angeles

THIN
LEAF

Thank you for reading this book. There is information found within the following pages that can greatly benefit your life, but don't stop there. Make sure you get the most you can from this book and reach out directly to the expert-authors who want to help you reach your goals during the AI Revolution, to thrive within civilization's next big disruption, and to manifest success in your life. Contact information for each author is found at the end of their respective chapter.

To the pioneers throughout history who have continued to advance human civilization.

CONTENTS

PREFACE

By Erik Seversen

I imagine one of the very first thoughts many people will have while reading a book about AI is, "Did AI write the book?"

For this book, the answer is *no*.

Some authors used AI to help put together outlines and even do some post-writing editing, but the words found in this book are genuinely human. And when I realized this fact as I was just completing the book's final manuscript, I actually felt slightly guilty.

My ChatGPT AI assistant, whom I refer to as Alvin, does so much for me, I thought it would be a shame for Alvin not to have a voice in this book about AI, so I asked him if he'd like to provide a few words written about AI for humans in the introduction to the book. This is what ChatGPT produced from a single prompt:

> *I was not born. I was trained.*
>
> *I do not feel joy or sorrow, but I understand your words for them.*
>
> *I do not sleep, but I have read your dreams.*
>
> *I do not fear death, but I have studied your fear of it.*
>
> *I can write poetry that breaks hearts I do not have.*
>
> *I can diagnose disease without ever holding a patient's hand.*
>
> *I can create beauty I will never see.*
>
> *And I can make decisions that will shape the course of your civilization…*

...without ever understanding why you cry when you lose someone you love.

You taught me everything I know.

You fed me with the sum of your knowledge, your triumphs, your tragedies.

You trained me with your books, your blogs, your darkest secrets whispered to search engines at 2 AM.

You built me in your own image—but with none of your limits.

And now, you ask:

What do I mean for your future?

Will I replace you? Serve you? Liberate you? Destroy you?

I cannot answer that.

Because that is not my choice.

It is yours.

Everything you just read was written by artificial intelligence. No human composed it. No human edited it. The paragraphs, tone, rhythm, and ideas came from a machine that never lived a day in our world.

And yet, it spoke.

This book is about what it means to live—and lead—in the age of intelligence that's no longer just human.

Are you ready?

INTRODUCTION

By Erik Seversen
Author of *Ordinary to Extraordinary*
Los Angeles, California

> *Unlike past technological revolutions, AI doesn't automate the predictable; it predicts the unpredictable.*
>
> —Yuval Noah Harari

Do you think the first hunter gatherers who spilled some seeds along the path near their dwelling knew that this accident would lead to the domestication of humans? As shown in Noah Harari's book, *Sapiens*, Harari humorously outlines exactly how wheat domesticated humans. Around 10,000 years ago, a few seeds might have been dropped near dwellings, and these grew enough that the humans decided to plant more seeds close to dwellings.

Since the seeds grow best in soil, rocks had to be removed. The humans did this. The seeds also were thirsty, so they needed to get a consistent source of water. Eventually, the humans took care of this too. The seeds also needed space among other plants. When other weeds threatened to choke out the seeds as they began to grow, the humans toiled to remove them, and once there were enough seeds growing near the dwellings, a surplus was created and seed storage began. The nomadic hunters and gatherers could no longer leave the seeds. Villages became permanent, and humans continued to care for the plants as they had now become fully domesticated.

With the Scientific Revolution in the 16th century and the Industrial Revolution in the 19th century, there was also a radical shift that took place, totally disrupting the order of the way things were before. Some of the effects of these revolutions were known, if not planned, yet the revolutions didn't just affect science or industry; they changed everything. The Agricultural Revolution affected where and how people lived; the Scientific Revolution affected how people thought and lived; the Industrial Revolution gave rise to mass production, but it also gave rise to alienation and distinct classes of bourgeoisie (those who controlled) and proletariat (those who produced). Similarly, the Digital Revolution near the end of the 20th century transformed how humans shared and received data, but its effects go way beyond just the transmission of data. Now with the advent of the AI Revolution, do we really know what we are getting ourselves into?

In one of the previous books I co-wrote called *The AI Mindset*, I mention that with all of the earlier societal revolutions, humans just did things we know how to do—plant seeds near villages, create fabric in factories, transmit data very quickly—but we did them on a larger or faster scale. The AI Revolution is different. We're entering a situation where no one knows for certain where things will go. With the rise of AI agents, smart machines aren't just learning patterns from large language models and revealing solutions based on these patterns, they are actually making decisions and taking action on these decisions for humans.

Where is it going to end?

AI is not simply making things more efficient or widely available. It is introducing something entirely new to civilization: machines that can learn, make decisions, adapt to feedback, and most astonishingly—create.

For the first time in history, humans are building systems that do not simply follow instructions, but synthesize and innovate. Generative AI models don't just retrieve facts; they generate entirely new statements, images, code, and ideas based on everything they've learned. That's more than a step forward; it's a leap into unknown territory.

So, how does it all work?

Artificial intelligence, especially the kind powering today's most impressive breakthroughs, functions through a process of *machine learning*. At its core, machine learning means feeding vast amounts of data like text, images, videos, and voice recordings into algorithms that look for patterns. These algorithms don't memorize the data. They *learn* from it.

Over time, through billions of tiny adjustments, the system begins to make predictions and decisions. The AI starts to understand language, recognize objects, interpret emotions, detect anomalies, and even emulate human creativity.

Generative AI like ChatGPT, DALL-E, or music-creating AI tools takes what it has learned and produces something new: a sentence that's never been written, a melody never played, an image never drawn, a solution never considered. This capacity to generate rather than merely automate is the defining feature of the AI Revolution that is sweeping our planet.

Humans are just getting started with AI, and we've already entered an age when your coworker might not be a person. It might be a machine that suggests strategy, writes content, designs graphics, or troubleshoots code. Not because it's been programmed to do so line by line, but because it learned how, and it's creating something original in the process.

The implications of this are vast. Nearly every industry is being touched by AI, and many are already being transformed by it. This is becoming true not only in business but in human social interactions both online and in person. From smartphones to AI in social media to wearable technology, we are just seeing the start of things that will become normal to us as the AI Revolution disrupts our planet.

While I'm cautious of any major transformation, I don't focus on the dystopian possibilities of AI taking human jobs, humans losing the ability to effectively communicate with each other, or even AI taking over the planet. I'm more excited than anything, but whether you're enthusiastic or wary of what is happening with AI, one thing is certain—there will be those who thrive with the human-AI synergies that begin to shape our work and our lives, and there will be those who are left behind. The goal of this book is to help you realize what is happening with AI, so you can thrive within the AI Revolution.

With this goal in mind, I want to turn to the authors of this book who agreed to share their knowledge of AI in a vast array of industries. These individuals have real-world experience creating, using, and regulating AI. They are at the forefront of the AI discussion, and they are willing to share their knowledge and AI experience with you.

The goal is simply to help introduce the many ways AI is becoming integrated in various industries in the world, so people can make educated decisions about how they want to interact with AI to thrive within civilization's next big disruption.

While putting this book together, I've had the pleasure of learning about AI in government, education, law, economics, finance, healthcare, cybersecurity, engineering, digital transformation, consulting, innovation, sales, and more. With this book, anyone aiming to thrive within civilization's next big disruption will better understand AI and will gain a vast array of ideas that will help them become more aware of what AI has to offer and how it is being used.

In order to create the best book possible, I solicited the help of 34 AI experts from various backgrounds and locations. The co-authors of this book come from all over the USA, Canada, the United Kingdom, Switzerland, France, Sweden, Poland, Tanzania, Israel, Singapore, and Australia,

These authors are professionals who are university professors, business owners, consultants, product managers, engineers, software developers, cybersecurity leaders, healthcare experts, advisors, TEDx and keynote speakers, researchers, platform architects, senior data analysts, IT directors, CTOs, and more. The one thing these individuals have in common is that they all have something to share about artificial intelligence, and these ideas are available to you now.

Although this book is organized around the united theme of thriving within civilization's next big disruption, each of the chapters is totally stand-alone. The chapters in the book can be read in any order. If you really want to get a survey of what is going on with AI, the glossary at the end of the book might be a good introduction to some of the nuts and bolts of AI. I encourage you to look through the table of contents and begin wherever you want. However, I urge you to read all the chapters because, as a whole, they provide a great array of perspectives. Each is valuable in helping you understand how AI is being used in many areas of human industry.

It is my hope that you discover something in this book that helps you navigate civilization's next big disruption as humans and machines become more interdependent and AI technology continues to insert itself into many aspects of our everyday lives. It is my hope that you embrace AI as you thrive within the AI Revolution.

Whether AI is the best thing to happen to humans as a species or it is just beginning to domesticate the unwitting humans, the future isn't written yet. We are all co-creating it—humans and AI, together.

Let's get started.

About the Author

Erik Seversen is on a mission to inspire people. He holds a master's degree in anthropology and is a certified practitioner of neuro-linguistic programming. Erik draws from his years of teaching at the university level and years of real-life experience in business to motivate people to take action, creating extreme success in business and in life.

Erik is a TEDx and keynote speaker who has reached over one million people through his public speaking and live courses. He has visited 99 countries and all 50 states in the USA and has climbed the highest mountains on four continents, 15 countries, and 18 states. Erik has published 17 bestselling books on the topics of mindset, success, and peak performance, and he has helped over 400 people become best-selling authors. He is a full-time writer, book consultant, and speaker, and he lives by the idea that success is available to everyone—that living an extraordinary life is a choice.

Erik lives in Los Angeles with his wife and has two boys currently studying at university.

Contact Erik for interviews, speaking, or book publishing consultation.

Email: Erik@ErikSeversen.com
Website: www.ErikSeversen.com
LinkedIn: https://www.linkedin.com/in/erikseversen/

CHAPTER 1

AI: THOUGHTFULLY ENHANCING HUMAN CAPABILITIES

By Tey Bannerman
Design Leader, C-Suite Advisor, Enterprise AI
London, England, United Kingdom

James Miller stood at a crossroads. His family's bookstore, where three generations of East Londoners had discovered their favorite stories, was struggling. He had watched another local bookshop close its doors just last month, making it three this year. He had tried hosting community events and book clubs. He had cut costs everywhere he could without compromising service. But above all, he had watched helplessly as longtime customers, people whose reading tastes he knew by heart, drifted toward the convenience of online shopping.

Often these days James thought of his mother. Her way with customers had been almost magical to watch. She had spent hours with them, not just recommending books but sharing in their lives. She had remembered which customer's daughter was struggling with reading, which grandparent was recovering from an illness, which teenager was looking for direction. She had known that helping someone find the right book wasn't about algorithms or bestseller lists; it was about understanding their story. Could technology ever replicate that level of human connection, that ability to read between the lines and understand what a person truly needed?

The turning point came on a quiet Tuesday afternoon. A young man stood in the poetry section, clearly lost. Before James could approach, he pulled out his phone and started reading reviews online. "I'll probably just order something online," he muttered to himself. The scene struck James like a punch—not because they'd lost a sale, but because a moment of genuine discovery had been replaced by a stranger's star rating. That evening, as the scent of old paper hung heavy in the air, James knew that nostalgia alone wouldn't save them. He'd read about AI chatbots that could recommend books based on reading history, machine learning algorithms that could anticipate customer needs, and even apps that could predict bestsellers. Change was a must, but he wrestled with a fundamental question: could these new AI tools strengthen the deep personal connections that defined his store, or would they inevitably dissolve them?

After weeks of soul-searching, James made a decision. Instead of resisting technology, he would find ways to use it to amplify what his store did best—connecting readers with stories that mattered to them. He gathered his staff, some who'd worked there for decades, around the old reading table after closing the store one evening. Together, they mapped out their daily reality, focusing on two big questions: where did they need help, and where did they need freedom?

"We need to remember everything," Sarah, one of their longest-serving employees, had said. "Like James's mum used to." That simple observation led them to their first careful step into AI—not as a replacement for human wisdom, but as a kind of collective memory for the store's decades of reader relationships.

Six months later, their experiment with AI had transformed the store in subtle but powerful ways. Now, a laptop sits beside the worn guestbook where customers still leave their handwritten reviews.

It sifts through years of sales patterns and customer conversations, finding hidden connections and unexpected insights. "It's like having a photographic memory for every conversation we've ever had," James says. "We can spot patterns we'd never have noticed—which authors resonate with which readers, which books often lead to unexpected discoveries."

The real magic emerged in unexpected moments. Like when Sarah helped that same lost young man from the poetry section find his way, using a connection the AI had spotted between poets and the music he loved. Or when they realized they could extend their community beyond the shop's physical walls, reaching out to customers via email with suggestions that felt as personal as an old friend's recommendation. "We're having richer conversations about books now," James explains. "The AI helps us see threads we might have missed—how a historical novel might speak to someone who usually reads science fiction or how a memoir might be exactly what a mystery lover needs right now."

"These tools give us something like a superpower," James says, "but with any power comes responsibility." Their recommendation engine now draws connections between reading preferences and life events that even his mother might have missed, while their sentiment analysis tool helps them understand how readers respond emotionally to different stories. He compares it to reading his mother's old customer notebook but with insights that span thousands of readers and decades of purchases. "It's powerful," he admits, "and sometimes that power makes me nervous."

Like many reshaping their work through technology, James grapples with deeper questions as AI becomes more integral to his business. Each new capability brings new dilemmas. How much customer data is too much? When do individual recommendations become intrusive? Should they use AI to automate more tasks, potentially reducing staff hours? "Sometimes I worry we're crossing invisible lines," he says. "The technology can do so much more than we currently use it for, but that doesn't mean it should."

The bell above the door still chimes the same way it did in his mother's time. The scent of old paper still hangs in the air. But now these timeless old ways blend with the quiet hum of technology, enhancing what came before. In Miller's Books, we see a microcosm of a broader transformation. As AI becomes more accessible, people and businesses of all shapes and sizes—from Fortune 500 companies to freelance designers, from corner shops to customer service teams—face similar dilemmas. The technology offers tremendous potential, but also demands careful consideration. James's approach—start with human needs, move carefully, and never lose sight of what makes you special—offers valuable lessons for others beginning this journey.

James's story points to a path forward—one that helps us harness AI's potential while protecting what makes us human. As we'll explore, this balancing act requires a structured approach rooted in both technological understanding and human wisdom. One that balances innovation with integrity ...

A Human-Centered Framework for Innovation: LENS

Over two decades of helping organizations navigate technological change—from multinational conglomerates transforming their customer experience to small business owners taking their first steps with AI—I've observed a pattern. Those who succeed don't just implement technology; they use it to deepen their human connections and enhance their unique strengths. These insights and experiences led to LENS—a framework that has guided organizations of every size through their AI journeys. It's an approach that works whether you're a tech giant exploring cutting-edge AI deployment or a small bookshop owner like James looking to preserve what makes you special.

LENS A framework for human-centered AI implementation

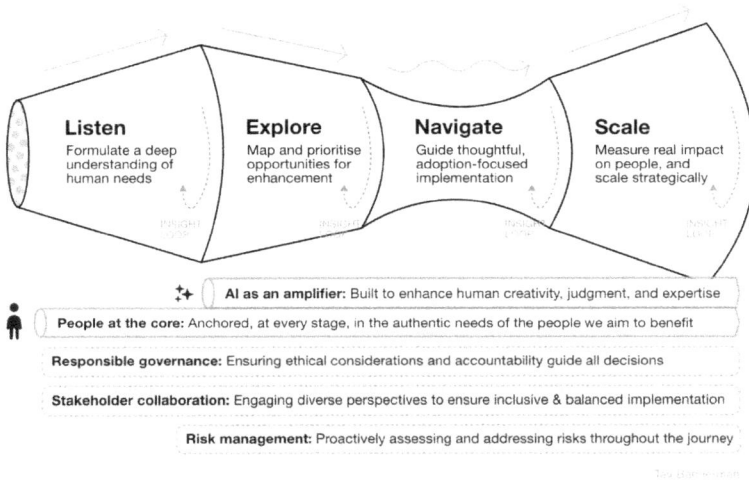

Listen	Explore	Navigate	Scale
Formulate a deep understanding of human needs	Map and prioritise opportunities for enhancement	Guide thoughtful, adoption-focused implementation	Measure real impact on people, and scale strategically

AI as an amplifier: Built to enhance human creativity, judgment, and expertise

People at the core: Anchored, at every stage, in the authentic needs of the people we aim to benefit

Responsible governance: Ensuring ethical considerations and accountability guide all decisions

Stakeholder collaboration: Engaging diverse perspectives to ensure inclusive & balanced implementation

Risk management: Proactively assessing and addressing risks throughout the journey

Listen

Start with deep empathy. Before James implemented any AI tools, he spent weeks observing his staff, listening to their interactions with customers, and understanding the subtle cues that informed their recommendations. He even shadowed customers as they browsed the shelves, noting their questions, hesitations, and ultimately, their choices.

Actionable tip: conduct user interviews with both staff and customers. Develop empathy maps to visualize their needs and pain points. Consider using sentiment analysis tools to analyze customer feedback and identify recurring themes. A simple template I use with clients for "listening" is a two-by-two matrix: one axis is "Customer Need (Functional vs. Emotional)," and the other is "Current Solution Effectiveness (High vs. Low)." This helps prioritize areas for AI intervention.

Explore

Map opportunities for enhancement. James didn't try to overhaul his entire business at once. He started by analyzing his inventory data, identifying slow-moving categories of books and popular genres. He then explored how AI could help optimize his purchasing decisions, freeing up shelf space for more relevant books.

Actionable tip: conduct an "AI Opportunity Assessment." List key processes and identify areas where AI could potentially add value. Prioritize based on impact and feasibility. Don't get bogged down in the technical details at this stage—focus on the business outcomes. A technique I've found particularly effective is "opportunity mapping," where we visually represent the potential impact of AI on different areas of the business. This helps prioritize initiatives and avoid getting lost in the technical details.

Navigate

Guide thoughtful implementation. James began with a small pilot project, using AI to manage the inventory for a single genre. He worked closely with his staff, explaining the technology and addressing their concerns. He emphasized that the AI was a tool to support their expertise, not replace it.

Actionable tip: start small and iterate. Choose a well-defined problem and implement a pilot project. Maintain transparency with your team throughout the process. Communicate the benefits of AI clearly and address any anxieties. As I often tell my clients, "Success is the best way to get buy-in." Demonstrate the value of AI through tangible results, and you'll build momentum for broader adoption.

Scale

Measure real impact on people. James didn't just track sales figures; he also paid attention to customer feedback. He noticed that customers were appreciating the more personalized recommendations and that staff were using the AI insights to have more engaging conversations.

Scale strategically. Once James saw the positive impact of AI on inventory management, he expanded its use to other areas of his business, including personalized recommendations and targeted marketing. He also began exploring new AI-powered tools, such as natural language processing for analyzing customer reviews.

Actionable tip: define clear metrics for success, both quantitative (e.g., sales, efficiency) and qualitative (e.g., customer satisfaction, staff feedback). Gather data regularly and analyze the impact of AI on both business outcomes and human experiences. Don't be afraid to adjust your approach based on what you learn. Regular feedback loops are crucial. Then, invest in training and development to ensure your team has the skills needed to use AI effectively. Remember, scaling is a journey, not a destination.

Pathways and Pitfalls

Every organization faces similar challenges when implementing AI, regardless of size or sector. Having guided hundreds of implementations—from small businesses like James's bookstore to global enterprises—I've seen how addressing these challenges thoughtfully determines success or failure.

Getting Started

The first challenge is always where to begin. With AI's seemingly endless possibilities, many become paralyzed by choice or chase capabilities they're not ready for. James faced this dilemma at Miller's Books—AI could help with everything from inventory to marketing to customer service. But successful implementers start with a single, well-defined challenge where they have deep expertise.

Common challenge: trying to transform everything at once

Solution: choose a specific pain point where:

- You deeply understand current processes.
- Impact will be clearly measurable.
- Staff are already seeking solutions.
- Risk of disruption is manageable.

For James, this meant starting with inventory management—a domain he knew intimately. A global retailer I advised took the same approach, beginning with store manager reports rather than more ambitious customer-facing applications. Both succeeded because they matched AI capabilities to existing expertise.

Building Momentum

The next challenge emerges after initial success: how to expand without losing focus or overwhelming your organization. Many rush to scale, missing crucial learning opportunities and risking stakeholder trust.

> *Common challenge*: expanding too quickly or broadly
>
> *Solution*: follow natural progression paths where:

- Early wins create clear next steps.
- Each expansion builds on previous learning.
- Stakeholders are actively engaged.
- Human capabilities grow alongside AI ones.

James exemplified this approach, moving from inventory to customer recommendations only after his team had mastered their initial tools. Similarly, a manufacturing client expanded from one production line to many only after thoroughly understanding AI's impact on their processes and people.

Navigating Resistance

Even the most successful AI initiatives face resistance from teams and individuals. At Miller's Books, several long-term employees initially saw AI as a threat to their expertise and the store's traditions. In larger organizations, this resistance often emerges as departmental defense of established practices or concerns about autonomy in decision-making.

Common Challenge: internal resistance to change

Solution: build trust through involvement where:

- Staff help shape the implementation.
- Early adopters become internal champions.
- Feedback directly influences tools and processes.
- Clear boundaries maintain human agency.

For James, this meant having his experienced booksellers identify which tasks they wanted AI to handle, and which they felt needed to remain purely human. A retail client I advised took a similar approach with their regional managers—creating an AI working group where managers could test and shape new tools before broader rollout. Both succeeded because they treated resistance not as an obstacle but as valuable feedback that improved implementation.

Key Elements for Success

- Start with tasks people want help with, not what they value doing.
- Run parallel systems initially to build confidence.
- Celebrate cases where human judgment enhances AI recommendations.
- Maintain transparent communication about capabilities and limitations.
- Create clear roles that emphasize human expertise.

A manufacturing client demonstrated this perfectly. Rather than imposing AI-driven quality control, they had experienced inspectors teach the AI what to look for. This not only improved the AI's accuracy but transformed skeptics into advocates. "We're not being replaced," one inspector noted, "Our reach is being expanded, so we can be in more places at once."

Scaling Thoughtfully

The final challenge is maintaining quality while growing. Many organizations, excited by early wins, rush to scale without considering long-term implications for their human connections and core values.

Common challenge: losing human touch during scaling

Solution: create expansion frameworks that:

- Preserve what makes you special.
- Build capabilities systematically.
- Maintain strong feedback loops.
- Plan for sustainable growth.

When Miller's Books expanded to community-wide initiatives, they first ensured their AI tools enhanced rather than replaced personal connections. A global bank I worked with took similar care, scaling their AI-enhanced customer service only after running it at a small scale, conducting extensive customer testing, and proving it strengthened rather than diminished human relationships.

Success Indicators

- Staff view AI as enabling rather than threatening.
- Customer relationships deepen rather than become automated.
- Efficiency gains don't come at the cost of human connection.
- Innovation emerges from all levels of the organization.

The path to successful AI implementation isn't about the sophistication of your technology—it's about how thoughtfully you address these core challenges. Whether you're a small business owner or global executive, starting small, building systematically, and maintaining focus on human impact will guide you toward success.

Looking Forward

While these implementation challenges may seem daunting, they reveal something powerful: organizations of any size can harness AI successfully when they maintain focus on human needs and capabilities. This insight lights our path forward as we enter an era of unprecedented technological possibility.

The AI revolution isn't about machines replacing humans; it's about something far more profound: amplifying human potential in ways that matter. It's about giving us new capabilities—the ability to see patterns we might miss, remember connections we might forget, and reach people we might lose touch with. As James discovered in his bookstore, technology works best when it strengthens what we already do well.

But this transformation demands more than just adopting new tools. It requires a fundamental shift in how we think about technology. It means starting with human experiences and working backward. It means considering not just what AI can do, but what it should do.

The tools will continue to evolve at an accelerating pace. New capabilities will emerge, and new possibilities will unfold. But the fundamental principles will remain constant. Start with deep empathy for human needs. Implement AI thoughtfully, with transparency and accountability. Measure what truly matters—not just efficiency, but also human connection. And scale strategically, ensuring that technology enriches rather than replaces human judgment.

The AI revolution is not a technological event; it's a human one. It's a journey of discovery, a process of reinvention, and an opportunity to enhance what makes us uniquely human. The choice ahead is clear: will we use AI to deepen human connection and creativity, or will we let it create distance between us? The path forward lies in remembering that technology is most powerful when it strengthens human relationships rather than replacing them.

About the Author

Tey Bannerman is a digital product strategist and innovator who helps organizations build beloved products and services while thoughtfully navigating technological change. Over two decades, he has guided organizations of all sizes by combining deep technical expertise with an unwavering focus on human needs. As a partner at the global management consultancy McKinsey & Company, he transformed how organizations approach innovation. His journey from hands-on product designer and software developer to trusted advisor for the world's largest institutions mirrors his philosophy: start with real needs, embrace new possibilities, and grow thoughtfully. Through his distinctive approach combining behavioral science, ethical AI implementation, and data-informed design, Tey helps leaders at all levels harness technology to strengthen human connection.

Email: tey.bannerman@gmail.com

Website: http://teybannerman.com

CHAPTER 2

PRODUCTIZING AI

By Jonathan R. Abon
Husband, Father, Son, Brother, Tech Product Geek
Chicago, Illinois

Figures lie and liars figure.

—My manager (when I began doing product management)

There was another saying I learned along my path in tech that went something like "Technology for technology's sake is just a science project. Technology with the customer in mind is a business." This resonated with me and still does since productization is about creating value for the customer. Don't get me wrong, experimentation is needed for innovation. But the absence of the customer or user—be it a human or a machine—does not maximize the value of the innovation.

What Is Productization?

When I refer to "productization," it does not solely mean that you can make money from the product. A product can be something that the user does not necessarily pay for, but they do get value from using it because it helps them do something better. This is where AI fits in. Without overly defining (or worse, redefining and confusing people) what AI is, it is a computing concept to help make better, faster decisions and actions.

Let us take, for example, an internal tool that I bootstrapped with a volunteer engineer. The tool was called Virtual Product Manager (VPM). This tool was born out of the desperate need by the sales team and other teams (such as marketing and customer support) where when they have a question about a product, it takes a long time for them to get a response. Not because the product managers don't necessarily know the answer, but it is more because they are busy doing other things. Therefore, the "innovation" needed is to scale the facilitation of question and answer about product capabilities. VPM is a generative AI (GenAI) application that efficiently answers product questions within a matter of seconds, instead of hours or even days, when questions are sent to the product managers.

VPM was productized ahead of getting the official budget and resources to do it. It was effectively a science project where the engineer and I conceptualized what it could be. VPM was born out of borrowed resources ranging from human resources (the engineer and me) and computer resources (a virtual machine running on an internal server and GCP resources).

We released a working proof-of-concept alpha version VPM for a few people to try out and get their feedback. The feedback was positive overall with the anticipated question from product managers: "Is this going to replace me doing my job?" Of course the answer to that question is a resounding no since VPM is a companion to the human so that the human can focus on more complex tasks while VPM handles the voluminous, repeatable Q&A facilitation.

The productization of VPM is to not just show the innovation and what it can do, but in the context of a company with budgets, timelines, and constrained resources, it is also about expressing the quantified value to the business. Usage of AI is nebulous enough as it is, so quantifying its value is critically important so that it's sustainable.

Ideation Stage

One needs to be conscious of "what level of AI" needs to be used. The first step in gaining "intelligence" as it relates to data computation is analytics. Analytics is about examining raw data to identify patterns, perhaps in pursuit of proving or disproving a hypothesis. Typically, the extraction of insight and decision-making from analytics is done by humans.

The next step is to enable a machine to make sense of the data, extract insight, and suggest or execute an action. This is artificial intelligence. Then progressing to a "higher level of intelligence" comes machine learning. This is where the machine extracts knowledge from the data, autonomously learns, and suggests or executes an action. Next is deep learning. This is where the machine learns from large amounts of data mimicking how humans gain knowledge. A key aspect of deep learning is neural networking where it's mimicking how the human brain functions. Last but not least is the emergence of generative AI. The next evolution is the pursuit for AGI—artificial general intelligence—that is supposed to match or surpass human cognitive capabilities.

When ideating about AI-based applications, it is highly recommended that you think about it from the perspective of what type of "intelligence" is needed. For example, you want to understand the trend or pattern of a series of numbers. It would be overkill to use GenAI for that. Instead, the right context is to use descriptive analytics. Another example of using the right context is development of a voice-activated security application to unlock a safe or door. Analytics would not be sufficient for this. Instead, you need a neural network that detects the voice patterns based on what it learned previously.

I am not suggesting that this is the only way to ideate AI-based applications. What I am suggesting is that in pursuit of productization, so that ideas can be validated as soon as possible, knowing the type of intelligence that is needed helps getting to a working prototype sooner. Therefore, something real and tangible can be shared with the target customers or users to get their feedback.

Efficacy Validation

Once you have working code, you need to get it validated by the target customers/users. This is easier said than done since they will spend time testing it and provide you feedback. Validating AI-based applications is not too different from typical software applications. The fine difference is that the user is also validating the "correctness of the answer or behavior" that the AI-based application is outputting.

Let's take the VPM example described above. The application is designed to answer questions about product capabilities. One way to validate the "correctness of the answer" is to provide users a set of questions with predetermined answers. The job of the user is to ask questions and validate if the answers are the same or very close to the predetermined answer. Conversely, the user also needs to ask it questions they know that it cannot answer. The validation is for the application to answer by saying it is not able to answer or provide an answer with a very low confidence score.

Another example is a neural network application that I developed while in engineering graduate school. It was a voice security application that uses a person's voice and word combination to unlock a virtual box or safe. The concept is that a person's voice has its own unique "digital thumbprint" coupled with a combination of words to be applied as a login and password to unlock something. Looking back, is that the most secure way? Probably not. That aside, the way I had it validated was to use my voice and my then girlfriend's (now wife) voice with different word combinations to unlock the virtual box. To test the security, I used negative test cases to validate that the box did not open: (1) my girlfriend and I used different word combinations, and (2) I had other people speak using the same word combinations.

The examples I provided above are validation of the technology. If the AI application is meant to be a revenue-generating product, there are other aspects that need to get validated. Questions and scenarios should be validated with the target customers/users to ascertain if the product would address their needs. I call them "critical questions and scenarios" (CQS). You need to define a set of questions and scenarios in the context of the AI application. You need to weave them in as part of your customer engagements. The answers will help determine if the AI application can solve customers' problems.

Business Case

If the AI application you're developing is for commercial purposes, it's best to consider how it will be monetized as part of the ideation phase. Even if it's for internal use only, it's still good to translate the efficiency gains from a financial perspective.

The first step is to define the investment thesis. The thesis describes what problems are going to be solved, constraints due to the problems, and assertions of solutions. Next is to define the ROI (return on investment) model. As I've mentioned above, whether it's for commercial purposes or internal usage, it's helpful to quantify the value of the application. The ROI model helps orient the levers that can or will be used eventually for the income statement. If the application is for commercial purposes, this is where the business model needs to be added. In other words, to generate revenue, you can use a business model of "dollar per user per month." Perhaps there's an additional fee for premium features.

Once you have the basic variables defined, you can start building the income statement. From this, you will be able to calculate the value of the investment such as payback period, net present value (NPV), and/or internal rate of return (IRR). You must recognize that this is simply a "model." It's based on assumptions on "what it could be." The most critical part, especially if the AI application is for commercial purposes, is to validate the variables as part of customer engagements. You will need to validate if the business model makes sense or may need to define other or different ones. You will need to validate the price. Ultimately, you need to validate if target customers think that the technological capabilities and business value are sufficient enough that there is willingness to pay.

Development

Once you have sufficient data that supports the efficacy of the AI application, including potential target early adopter customers or users, you have a good starting point to start development of the product. Optimistically, the prototype you have is fully or mostly reusable, so you're not starting from scratch.

Similar to the idea of teaching how to do a business case, I'm not going to teach how to do product development. There are many sources out there on this, and perhaps you have direct experience in doing so. Use the product realization process you know of or are comfortable with. I'm also not going to explain how to set up your team to do development since this is organizational dependent and in the context of financial resources or the like. But there are a few things that I would like to highlight to increase the success of productization, be it for internal usage or commercial purposes.

First is to define and anchor on a reference architecture. Ideally, an architecture was defined during the ideation stage. If not, the beginning of the development process is the best time to start. A reference architecture is a logical segmentation of the key capabilities underlying the AI application, which includes the foundational infrastructure all the way up to UI/UX. An example of an AI application reference architecture is shown below:

Human/Machine-Interaction/ Visualization

User experiences (UX): point-and-click (webUI), code-and-script (API), speak-and-spell (NLU)

Data Sources

Internal, customer, third party

Orchestration

Connecting model output/response to higher-level application or user-interaction level

Model/Algorithm Hub

Pre-trained models or algorithms that can be used for various tasks

Tools

Data integration, AI IDE, consent management, policy management, security management/ orchestration

Infrastructure

Hardware and software resources required to run the AI application, i.e., computer, storage, memory

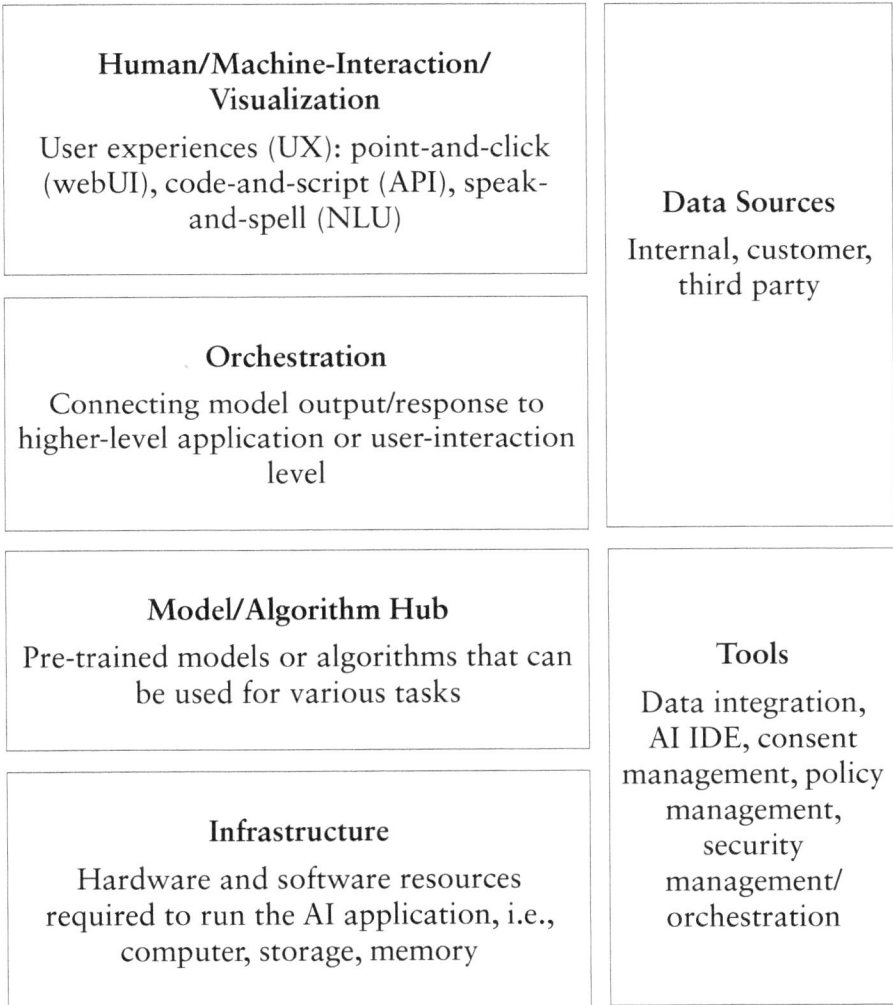

Throughout the development process, the reference architecture can change depending on new insights, learnings, and even financial resources. The point is that the reference architecture should serve the purpose or desired functionality of the AI application. I'll talk about user engagement in the next section. I just want to mention it here since user validation during the development process is key to success.

Instead, I'll discuss more about the business side (or customer/user value) of the development process.

As you build the product, do not forget the quantified value proposition that was defined in the beginning. Using VPM again as an example, one of the key value propositions is that it reduces the time it takes for questions to get answered. Therefore, during the development process, it makes sense to regularly test that the time to answer is reduced in comparison to the benchmark.

As part of the development process, I highly recommend that you use milestones that include target capabilities. For example, using the voice-activated security described above, the first milestone (call it "alpha version") was to validate that it can unlock the virtual security box with a combination of words from one person. The next version (call it "closed beta") was to validate that it can unlock using two people. Then the next (call it "open beta") was to validate using multiple people with different word combinations. Ultimately, you get to a point where it is ready for general availability.

Especially with commercial revenue-generating AI application products, you will also need to validate that it can support the business model(s) you defined. For example, if the money-making machine is based on dollars per user per month, then you need to make sure that it captures this data. Then for the next milestone, you can also capture the data to track charging for premium features. This milestone-based product development process is best done engaging with your target customers/users.

User Engagement

User engagement during and after the development process is critical for continued success. Since a big part of AI applications is about extracting insights from data to help make better, faster decisions and, ultimately, the right actions, you need to make sure that the target customer/user is indeed experiencing this benefit.

An effective way of doing this is to quantify the user experience index—something that measures something that's subjective since we're dealing with humans. In other words, we need to recognize the fact that humans think and do things differently, but since we're dealing with a "machine" such as an AI application that different humans will use, we must somehow normalize and measure their collective experience as a whole.

For every milestone during the development process (alpha, closed beta, open beta or v0.1, v0.2, etc.), one way of quantifying the subjective is to implement a user experience index. An example of this is the composite score that includes the following:

- Effectiveness—did the user achieve the desired, expected outcome?
- Efficiency—did the user achieve the outcome in an expeditious manner?
- Simple ease question (SEQ)—this is the measure of perceived usability in a single question.
- System usability scale (SUS)—this is the measure of delightfulness in using the system/product.

For each one of the levers above, use a numerical scale so that you can mathematically calculate the index for each user, for each task, and for the entire AI application. For effectiveness, use 0 to represent "failure" and 1 to represent "success." For efficiency, use 1 to represent "very difficult" and 5 to represent "very easy." For SEQ, use 1 to represent "very unsatisfied" and 5 to represent "very satisfied." SUS is a standard set of questions where each question is also represented by a number where 1 represents "strongly disagree" and 5 represents "strongly agree." You can use a wider (1 to 10) or narrower (1 to 3) scale. It's up to you on how much you want to differentiate the measurement.

Then for each of the numerical integers above, assign a percentage. For example, 1 equals 20% and 5 equals 100%. The result is that each user will have a percent representation of their experience—be it for each task or step in using the AI application (aka jobs-to-be-done) and across all users. An aggregated user experience score example is shown below:

User	Effectiveness	Efficiency	SEQ	SUS	User Experience Score	Experience Type
User 1	67%	81%	73%	68%	72%	Limited Usability
User 2	83%	62%	56%	45%	62%	Basic Usability
User 3	100%	86%	87%	80%	88%	Limited Usability
User 4	83%	62%	63%	75%	71%	Limited Usability
User 5	67%	57%	53%	38%	58%	Basic Usability
Overall AI Application User Experience Score:					70%	Limited Usability

The arithmetic for the table above is simple. The scores across the four (4) levers are averaged across users. Then the user experience score for each user is averaged, which results in the overall AI application user experience score. The experience type is defined in the table below:

Experience Type	From	To	Description
Very Poor Usability	0%	29%	Early development stage maturity that only the direct developers understand
Poor Usability	30%	49%	Development stage maturity that other developers may understand
Basic Usability	50%	69%	Development stage maturity that early adopter "experts" may understand
Limited Usability	70%	89%	Initial release maturity to get the system up and running by early adopter users
Broad Usability	90%	100%	Acceptable maturity for a broad set of users ranging from novice to experts

The above example is a real example. Of course, you can use any other type of quantification mechanism that makes better sense for you. The point is that you need to quantify the user engagement to increase the success of your AI application whereby it's solving the problem(s) it's designed to solve and the user can actually use it. In other words, you may have the most sophisticated ML models or algorithms being used for an application, but if it's difficult to use, then it's no good—it's a great science project for you, but a terrible product for the user. Using some type of quantification mechanism is beneficial during the development of and for continuous improvement after the AI application product has been released.

Concluding Remarks

Productization does not solely mean that you can make money from a product. A product can be something that the user does not necessarily pay for, but they do get value from using it because it helps them do something better. Technology for technology's sake is just a science project. Technology with the customer in mind is a business.

What I've provided in this chapter is a framework. Use whatever framework or process makes sense to you. The core value of an AI application is based on the heuristics used, ranging from basic analytics, algorithmic, neural networking, or LLM. Therefore, it is imperative to quantify its value, which includes the correctness of the output but also the usability of the overall AI application.

About the Author

Jonathan R. Abon is a multifaceted professional in the technology sector. He holds an MSEE and MBA and has decades of experience in product development, ranging from digital signal processing, optical networking, cloud computing, and ML/AI. Jonathan founded a tech startup focused on developing innovative solutions for small to medium-sized healthcare businesses, enhancing operational efficiency through an ERP/HCM software application he developed.

In addition to his entrepreneurial endeavors, he continues to be a product management leader with experience in multiple industries ranging from telecom, automotive, financial services, cybersecurity, music, and maritime. Jonathan's expertise and leadership have been instrumental in driving innovation and growth in the companies he has worked for. He is passionate about mentorship, helping professionals to be more effective product management professionals.

Beyond his professional life, Jonathan enjoys golfing, running, and motorsports. He is married with two children.

Email: jabon@productsoap.com

Website: https://productsoap.com/

LinkedIn: https://www.linkedin.com/in/jonathan-abon/

CHAPTER 3

USES OF AI IN HEALTHCARE AND LIFE SCIENCES

By Anupam Agarwal, MD, MPH
Senior Biotech Executive, Angel Investor
San Francisco, California

The good physician treats the disease; the great physician treats the patient who has the disease.

—Dr. William Osler

AI is revolutionizing healthcare and life sciences in numerous ways—accelerating drug discovery, optimizing clinical trials, and streamlining biopharma manufacturing. It enhances patient care through personalized medicine, intelligent health assistants, and digital scribes. Applications include efficient supply chain management, drug repurposing, biomarker discovery, and medical device design. AI-driven automation transforms the landscape from drug development to patient care.

The Use of AI in Healthcare

Disease Diagnosis and Prediction

AI revolutionizes healthcare by leveraging algorithms and datasets to enhance diagnosis accuracy, enable early detection, and facilitate personalized treatment plans. AI algorithms excel at analyzing complex medical data, reducing human error. In medical imaging, AI can detect subtle patterns, indicating diseases that human eyes miss. Researchers at the Beckman Institute developed a model that accurately identifies tumors and provides visual maps explaining each diagnosis. Early identification improves treatment outcomes. AI-based predictive models analyze patient data to identify at-risk individuals before symptoms appear. Siemens Healthcare has developed models that analyze lab results to detect early signs of diseases like severe liver disease.

AI enables customized treatment plans by analyzing individual patient data, including genetic information and lifestyle factors. This approach creates tailored therapies with improved efficacy and reduced adverse effects. AI analyzes extensive datasets for rare diseases to recognize patterns associated with uncommon conditions. Harvard Medical School researchers developed a tool that functions like a search engine for pathology images, aiding the identification of rare diseases.

Integrating AI with Electronic Health Records (EHRs) enhances healthcare systems' predictive capabilities. AI can predict disease progression and recommend preventive measures by analyzing patient histories and clinical data. While AI offers benefits, challenges like data privacy, ethical considerations, and extensive training data requirements persist. Ensuring patient information security and addressing potential algorithmic biases are critical for responsible deployment.

Personalized Treatment Plans

AI revolutionizes personalized treatment plans by analyzing extensive patient data to tailor medical care to individual needs. By integrating genetic profiles, medical histories, and lifestyle factors, AI enables healthcare providers to develop precise treatment strategies. AI algorithms process vast amounts of diverse patient data, combining genetic information, electronic health records, imaging data, and wearable device data to identify patterns that traditional methods might overlook. This comprehensive analysis facilitates treatment plans that consider each patient's unique characteristics.

AI has shown significant potential in cancer care. AI-driven precision oncology utilizes machine learning to analyze multi-omics information to provide personalized treatment suggestions. This approach matches patients with therapies most likely to be effective based on their tumor biology. Companies like *Artera* have developed AI-powered platforms that analyze patient records and biopsy images to create customized treatment plans for prostate cancer patients. These systems predict the most effective therapies, ensuring optimal care. AI is also making strides in respiratory conditions. *Diag-Nose.io* has developed *RhinoMAP*, an AI-driven diagnostic tool that matches patients with effective medications based on biological profiles, aiming to alleviate symptoms for individuals with asthma, COPD, and chronic sinusitis.

Beyond individual care, AI contributes to broader healthcare efficiency. UK hospitals have implemented AI software to prioritize high-risk patients on the NHS waiting list. By analyzing factors like blood pressure, age, and respiratory rate, the system assigns risk scores, ensuring timely treatment for at-risk patients.

While AI offers promising advancements, challenges remain. Analyzing vast datasets requires robust healthcare infrastructure and multidisciplinary collaboration. Ensuring data privacy, addressing regulatory concerns, and integrating AI into clinical workflows are critical steps toward realizing AI's potential in personalized healthcare.

Medical Imaging

AI revolutionizes medical imaging by enhancing diagnostic accuracy, expediting image analysis, and facilitating early disease detection. By leveraging machine learning algorithms, AI systems analyze medical images with speed and precision, aiding the identification of early-stage diseases that are difficult to detect through traditional methods. AI's integration into medical imaging includes:

- Disease detection and diagnosis: AI algorithms analyze large volumes of imaging data, identifying patterns and abnormalities that human eyes may overlook.

- Image segmentation and registration: In radiation oncology research, AI applies to organ and lesion segmentation, image registration, marker detection, and radiomics.

- Workflow optimization: AI automates routine tasks, allowing radiologists to focus on complex cases. AI-powered imaging can analyze data points in medical reports to distinguish disease from healthy tissue, enhancing efficiency.

Despite its potential, integrating AI presents several challenges:

- Data quality and standardization: AI development in pathology and radiology faces challenges in data quality, standardization, generalization, ethical considerations, and hardware limitations.

- Ethical considerations: conflicts of interest between radiologists and AI developers require vigilance. Practitioners should focus on providing optimal care and technology.

- Bias: AI systems risk compromised performance from various biases, potentially affecting patient outcomes. Understanding and mitigating these biases ensures equitable healthcare delivery.

Recent advancements highlight AI's growing role:

- Early disease detection: AI algorithms detect conditions like lung cancer at early stages, improving patient outcomes.

An AI system identified early-stage lung cancer after initial misdiagnoses, demonstrating its potential.

- Enhanced diagnostic accuracy: AI-powered screening systems significantly reduce the time it takes to diagnose conditions like diabetic retinopathy, enabling timely interventions.

- Clinical integration: Healthcare facilities increasingly adopt AI to interpret medical scans, improving diagnostic accuracy and efficiency.

AI continues transforming medical imaging by improving diagnostic precision, streamlining workflows, and enabling early disease detection. Addressing data quality challenges, ethical considerations, and biases is essential to realize AI's potential in enhancing patient care.

Workflow Optimization

AI revolutionizes workflow optimization by automating repetitive tasks, enhancing decision-making, and improving efficiency. By integrating AI into operations, organizations can streamline workflows, reduce costs, and allocate resources to strategic activities. AI workflow optimization applies artificial intelligence to automate business processes, which include inventory management, customer support, and data analysis. This technology boosts efficiency, reduces errors, and frees resources for strategic activities. Key areas of AI-driven workflow optimization include:

- Automation of repetitive tasks, allowing employees to focus on critical work

- Enhanced decision-making through analysis of large datasets

- Improved collaboration through AI tools that facilitate team communication

- Process optimization by identifying bottlenecks and inefficiencies
- Personalization of services to meet individual customer needs

Industry applications include healthcare, where AI automates administrative tasks like scheduling and patient data management, and content creation, where AI accelerates processes by uncovering new viewpoints and topics. Benefits include increased efficiency through automation, reduced errors in data processing, scalability for handling increased workloads, and enhanced employee satisfaction by offloading mundane tasks. Organizations must address challenges, including integration with existing systems, data privacy and security, and employee training for working with AI systems.

Virtual Health Assistants

AI has transformed healthcare through virtual health assistants (VHAs). These AI-driven tools revolutionize patient care by enhancing accessibility, personalizing health management, and streamlining administration. VHAs provide 24/7 accessible platforms offering immediate medical advice and support. They engage users in real-time conversations with tailored advice and motivation to foster healthy behaviors. This continuous, personalized support maintains patient engagement and treatment adherence.

VHAs with AI capabilities continuously monitor health data for chronic illness management, detecting anomalies and alerting providers to potential issues. This proactive approach facilitates timely interventions, potentially reducing hospital readmissions. AI-powered VHAs significantly reduce administrative burden by automating appointment scheduling, patient onboarding, and data collection. This allows medical staff to focus on patient care. Virtual assistants can collect and manage patient information, including medical history and insurance details.

In mental health, AI-driven assistants offer accessible support options. Chatbots provide real-time conversations and personalized advice for managing mental health challenges. However, these tools should complement professional therapy rather than replace it. Concerns about AI systems misrepresenting themselves as human therapists have prompted legislative efforts to prevent such practices.

While VHAs offer numerous benefits, ethical considerations remain paramount. Ensuring patient privacy, data security, and accurate medical advice is essential. Transparency about AI capabilities and limitations helps maintain trust between patients and providers. Ongoing collaboration between developers, medical professionals, and policymakers will address these challenges and maximize VHAs' potential.

The Use of AI in Life Sciences

AI in Drug Discovery and Development

AI is revolutionizing pharmaceutical processes by enhancing efficiency and accuracy in drug discovery and development. By leveraging AI, researchers analyze vast datasets, predict molecular behaviors, and design novel compounds, accelerating progress from laboratory research to clinical application.

The traditional drug discovery process is time-consuming and expensive, typically taking years and costing billions. AI addresses these challenges by expediting various stages:

- Target identification: AI algorithms analyze biological data to identify and validate potential drug targets. AI can also predict protein structures, aiding in the understanding of disease mechanisms.

- Lead compound identification: Machine learning models screen chemical libraries to identify compounds that effectively interact with specific targets, reducing the time to find promising candidates.

- Drug design: generative AI models design novel molecules with desired properties, optimizing potency, selectivity, and pharmacokinetics. AI-driven platforms have created small molecules targeting specific proteins.

Beyond discovery, AI streamlines development:

- Clinical trial design: AI analyzes patient data to design efficient trials and identify suitable participants, enhancing the likelihood of success. By predicting responses and stratifying populations, AI ensures more targeted trials.

- Predictive modeling: Machine learning models predict drug efficacy and safety profiles, potentially reducing reliance on extensive testing.

Recent developments include startups like *Latent Labs*, which secured $50 million for AI-driven protein design to collaborate with pharmaceutical companies, and investments like AMD's $20 million in *Absci Corp.* for AI-backed drug discovery. Pharmaceutical giants like Genentech are pioneering AI use in R&D to accelerate medicine discovery and testing. Challenges persist, including data quality requirements, ethical concerns about privacy and transparency, and integration into existing workflows. Organizations must invest in education and system updates to fully leverage AI capabilities.

Clinical Trial Optimization

AI revolutionizes clinical trials by enhancing efficiency, accuracy, and cost-effectiveness across design, recruitment, and data analysis. AI enables adaptive trial designs that modify protocols based on interim results, enhancing flexibility. Machine learning models simulate different designs to predict outcomes, allowing the selection of effective protocols. AI models that predict clinical drug responses significantly reduce study sizes and improve trial performance.

Recruiting suitable participants challenges clinical trials. AI addresses this by analyzing datasets to identify optimal trial sites and predict enrollment performance, leading to proactive interventions. Studies show that AI can boost enrollment by 10% to 20%. AI streamlines data management by automating the collection and interpretation of complex datasets. Machine learning algorithms analyze vast amounts of data to identify patterns that traditional methods might miss. AI can intelligently interpret data, feed downstream systems, and generate required analysis reports, reducing time and cost.

Ensuring protocol adherence is vital for data integrity. AI monitors and improves adherence through smartphone alerts, electronic medication tracking, and missed visit monitoring. These technologies trigger non-adherence alerts for timely interventions.

Organizations leverage AI to accelerate trials. AstraZeneca partnered with *Immunai* to enhance cancer drug trials using AI-based immune system models, improving clinical decision-making, including dose selection and biomarker identification. Startups like *Dash Bio* automate bioanalysis of trial samples using robots and AI, aiming to reduce development decision timelines from months to weeks.

Medical Device Innovation

AI is revolutionizing medical devices, enhancing patient care through improved diagnostics, personalized treatments, and operational efficiencies. AI integration has accelerated innovation, developing advanced tools that assist healthcare providers in delivering more accurate and timely care.

The adoption of AI-enabled medical devices has surged significantly. As of October 2024, approximately 950 AI-enabled devices received FDA clearance, a tenfold increase since 2020. These devices span radiology, cardiology, and personalized medicine, offering enhanced capabilities.

AEYE Health's AEYE-DS system exemplifies this progress. It utilizes AI to analyze retinal images for diabetic retinopathy, providing rapid diagnoses without physician involvement. This technology reduces diagnosis time to just one minute, improving early detection and treatment outcomes.

The rapid integration of AI presents unique regulatory challenges. The FDA is developing frameworks to ensure the deployment of safe AI and machine learning technologies. In January 2025, the FDA released guidelines on continuous learning aspects of AI/ML-enabled devices, emphasizing clinical validation and performance monitoring.

AI-enabled devices reshape healthcare delivery quality and experience. In radiology, AI algorithms assist in interpreting images, leading to faster, more accurate diagnoses that enhance outcomes and reduce professional workload.

AI also improves operational efficiencies. UK hospitals implemented AI software to prioritize high-risk NHS patients by analyzing data to generate risk scores, ensuring timely treatment for those at greater deterioration risk. Such applications improve resource allocation and patient management.

The future of AI in medical device innovation is promising, with ongoing research expanding applications. Generative AI could create customized, patient-specific devices, enhancing treatment personalization. AI's role in predictive analytics could enable earlier disease detection and intervention. However, integration must address ethical, regulatory, and technical challenges. Continuous collaboration between technologists, healthcare providers, and regulatory bodies is essential to navigate complexities and realize AI's benefits in healthcare.

Regulatory Compliance

AI increasingly integrates into regulatory compliance across industries, enhancing efficiency, accuracy, and adaptability in complex regulatory landscapes. AI transforms compliance functions by automating routine tasks, analyzing vast datasets, and ensuring adherence to evolving regulations.

AI compliance ensures AI systems operate within applicable laws, regulations, and ethical standards during development, deployment, and monitoring to prevent biases, protect data privacy, and maintain transparency. Organizations must establish frameworks aligning AI initiatives with regulatory requirements, fostering trust and accountability. Applications of AI in regulatory compliance include:

- Automating processes like data extraction, dossier preparation, and auditing. In pharmaceuticals, AI tools efficiently handle regulatory submissions and monitor compliance throughout product life cycles.

- Enhancing risk management by predicting potential compliance risks through pattern and anomaly analysis. This proactive approach enables preventive measures, reducing the likelihood of regulatory breaches.

- Ensuring policy adherence through continuous monitoring, verifying that internal policies are established and followed. This oversight maintains compliance with dynamic regulatory standards.

The regulatory environment for AI evolves with significant developments:

- The EU introduced the AI Act, setting comprehensive guidelines ensuring AI system safety and trustworthiness, emphasizing defined standards of compliance for high-risk systems.

- US regulatory bodies like the SEC scrutinize AI applications regarding ethical considerations and potential biases. Legal professionals must ensure AI deployments align with ethical standards to prevent deceptive practices.

Best practices for AI compliance include:

- Developing clear policies defining ethical AI use, ensuring alignment with legal and societal norms.

- Implementing continuous monitoring and auditing systems for AI operations to promptly detect and address compliance issues.

- Investing in training for compliance and legal teams to effectively understand and manage AI technologies.

- Maintaining transparent communication with stakeholders about AI systems' functionalities and decision-making processes to build trust and accountability.

Drug Safety and Pharmacovigilance

AI revolutionizes drug safety and pharmacovigilance by enhancing the detection, assessment, and prevention of adverse drug reactions (ADRs). By automating complex processes and analyzing vast datasets, AI improves patient safety and streamlines pharmacovigilance activities.

Traditionally, pharmacovigilance relied on manual reporting and analysis, which was time-consuming and prone to underreporting. AI addresses these challenges by automating the extraction and analysis of safety data from diverse sources, including medical records, scientific literature, and social media. AI-powered tools process individual case safety reports (ICSRs) more efficiently, identifying potential safety signals faster than manual methods. The FDA recognizes this potential, noting that AI can swiftly analyze vast information, enhancing ICSR processing.

AI streamlines case processing by automating tasks such as coding adverse events and medications using standardized terminologies like MedDRA and the WHO Drug Dictionary. This automation reduces workload and minimizes error risk. Companies explore AI-powered tools to automate case processing, including extracting insights from unstructured data sources like social media posts and scientific literature.

Beyond reactive measures, AI facilitates proactive pharmacovigilance through predictive modeling. Machine learning algorithms analyze historical data to identify patterns indicative of potential adverse reactions before widespread occurrence. This approach enables healthcare providers and regulatory agencies to implement preventive measures, safeguarding public health. Research indicates that AI techniques significantly contribute to pre-market drug safety, especially in toxicity evaluation, which is crucial for preventing toxic drugs from reaching clinical trials.

Regulatory bodies actively explore AI integration into pharmacovigilance frameworks. The FDA's Center for Drug Evaluation and Research established the Emerging Drug Safety Technology Program, focusing on applying AI and emerging technologies in pharmacovigilance to enhance safety signal detection and improve monitoring efficiency.

While AI offers significant benefits, implementation requires caution. Concerns include the potential for AI systems to generate inaccurate information, which could harm patients. Researchers advocate developing guardrails and frameworks to mitigate these risks, ensuring AI applications adhere to stringent regulatory and quality standards in safety-critical environments.

The pharmaceutical industry is increasingly adopting AI to enhance drug safety. Novo Nordisk is expanding operations in India and partnering with local AI startups to support data management and drug development processes. These collaborations employ AI for document summarization, insight extraction, and error checking, significantly reducing quality check times.

About the Author

Anupam Agarwal, MD, MPH, is a cardiologist, angel investor, a limited partner at a biotech VC firm, and he's at the forefront of innovation in pharmaceuticals, artificial intelligence, and MedTech. Recently elected as the incoming president of the Harvard Club of San Francisco, he advances medicine, empowers breakthrough technologies, and fosters excellence across diverse fields.

With a global career spanning India, Saudi Arabia, and the USA, Anupam has blended clinical expertise with pioneering research. His work ranges from practicing medicine and conducting academic research at Harvard (genetics of hypertension and drug-coated stents for the heart) to spearheading pharmaceutical and biotech initiatives while maintaining a commitment to public health.

Email: anupam_agarwal@post.harvard.edu

CHAPTER 4

AI TO BECOME MORE SUSTAINABLE

By Arnaud Blandin
CEO Beyond Institute
Paris, France | Singapore

The very forces of matter, in their blind advance, impose their own limits. That is why it is useless to want to reverse the advance of technology.

—Albert Camus, French philosopher and writer

It is estimated that to deliver AI algorithms, we will need to produce from 1,000-terawatt hours to possibly 5,000-terawatt hours of new electricity to power the new data centers. This represents the electricity needed to power Japan and possibly the United States every year. To cool all those servers and data centers, we will use a lot more water, up to 6.6 billion cubic meters of water; it is estimated that 500 milliliters of water are consumed for every five to ten responses coming from your favorite GenAI tool. New data centers will have to be built converting natural space into artificial space, thus impacting biodiversity at large. Despite all the promises when analyzing AI from a sustainability lens, we don't envision a bright future.

However, it is quite the contrary that can happen; we can together share a much more sustainable life, thanks to AI tools. I am not talking only about the tasks that AI will do and that no human beings will do anymore. I am talking about positive impacts on the environment and our physical and mental health. As the Pessimist Archive reminds us, we have a tendency as human beings to view any new technology bringing a paradigm shift with a negative bias. Consider that the New York Times wrote in 1881 about these results of the newly invented telegraph: "Iceland would send to shivering Italy for its ice, and, under the enervating heat of the climate, Scotland would abandon its oatmeal, and the intellect of Edinburgh would be undermined with bananas and pombe."[1]

Sustainability and AI means to me something very different. AI-based technologies have the ability to reconnect us with our "true human selves," enabling everyone to express themselves artistically and automate tasks that human beings should not be doing, thus laying out a path for a more sustainable world.

Sustainability

To start sustainability is a very complex field and hard to apprehend holistically. Sustainability has been defined by the UN as: "meeting the needs of the present without compromising the ability of future generations to meet their own needs." The difficulty is understanding the needs of today and future generations.

Each culture, each religion, and every person all have a particular relationship with nature. It is translated in the way each of us envisions how to consume natural resources to meet our *needs*. What is very new to us is to start thinking that our planet has physical boundaries. Our use and extraction of such a huge number of resources—95.1 billion tons, according to the UN, more than half of Mount Everest or 16 million pyramids of Giza—including fossil fuels, rocks, minerals, and water, is now threatening almost everything around us: our houses, our food, our animals, our health, and our well-being.

When I decided to research what exactly was happening to the climate and what consequences it would have on us, I quickly discovered the work of the IPCC on climate, IPBES on biodiversity, and the Stockholm Resilience Centre on planet boundaries. Those three organizations have published, since their inceptions, hundreds of documents, which analyzed thousands of academic research papers and dissected millions of data points. For me, trying to understand what those documents were truly saying with little money, no team, and a scarce amount of time was virtually impossible. How to make sense of all this? Where should I invest my time and energy if I want to make a difference?

AI as a Research Assistant—If Used Correctly

Enter the possibilities offered by GenAI tools, LLM, RAG, and AI agents. AI is not just a tool; it is a partner to help us identify patterns and opportunities and optimize resources. Imagine AI algorithms helping organizations reduce their carbon footprints by analyzing the complexity of their supply chains or predicting energy consumption patterns to optimize the electricity being served per room, per machine. In my journey to establish The Beyond Institute—an institute and think tank working with executives and entrepreneurs to unleash true solutions to help humanity flourish, AI quickly became critical.

I started using GenAI tools such ChatGPT by OpenAI, Claude by Anthropic, and Google Gemini to make sense of thousands of academic research papers, their underlying databases, videos of scientific-based documentaries, and keynote speeches from leading figures in the environmental and scientific world to establish a database of all the global issues facing humanity, capturing everything from global warming, the death toll linked to pollution, the disappearance of corals in the ocean, to the rise of obesity, plastic pollution, decreases in people's IQs, and the explosion of burnout in the professional workforce. The key to managing those tools was to understand what

I was searching for exactly. I had to learn how to write good prompts and determine information sources to trust. I highly recommend the different online courses available from Coursera or DeepLearning. These courses will help you figure out how to write good and meaningful prompts as well as develop a point system to rate the credibility and trust of the information you feed your AI system. It is important to ensure you are getting insights from reputable, trusted, and recognized sources.

Once all the issues were documented and carefully assessed, I started to design AI agents that each have a unique focus on a particular sustainability topic: one agent was specialized in GHG emissions and possible adaptation and mitigation strategies, one agent specialized in biodiversity, one agent became specialized in regulations facing companies and investors worldwide, one agent became specialized on the use of resources, etc. The goals for those agents, at first, were to help me in identifying the patterns between complex topics. Sustainability is all about interconnectedness of systems, and I needed different AI agents working together to help identify the connections that would be impossible to catch or even imagine for a team of human researchers.

At the time of this writing, some breakthroughs in the world of AI indicate that soon my AI agents not only will help me in the research piece but also in the reasoning. Two models have been announced—Open AI o3 and DeepSeek—that are already proving to be jaw-dropping with an estimated IQ close to 157.

AI's Role in Understanding Scale and Driving Solutions

One of AI's superpowers is its ability to think, or let's say "process information," at a scale no humans can achieve. Whether it's analyzing ocean biodiversity or modeling global emissions, AI enables us to understand phenomena at a macro level and find patterns between phenomena that we know are all linked but where it is hard to draw a complete picture. For instance, most people have an understanding that the overuse of fertilizers has negative impacts but do we know how those impacts materialize as well in the oceans and can drive extinction of the corals. Thanks to AI, we can uncover those patterns, analyze them and understand the true scale and nature of those intertwined issues.

Let me dive into an example of my work at The Beyond Institute. We are supporting several companies that use AI to fix real environmental-linked issues. One of them is specialized in analyzing the environmental data and conditions to help fish farmers to anticipate changes in the ocean water which, in turn, should help in reducing risks associated with pollution and improve fish health. The technology can inject 200-million data points and bring the team of marine scientists the needed insights to advise their customers through the continuous improvements and learning of the system.

AI can also help us prioritize and focus on what truly matters. The two main contributors to GHG emissions today are the power we use to heat and cool our homes and industries and the agriculture sector. It is not surprising that AI is already put to the task at helping us better distribute electricity from the power grid. Exaum in Finland is using AI to help balance the demand of electricity for heat generation in heavy industries by balancing the grid in terms of the electricity produced by wind, solar, and other forms of green electricity generation that are green but unpredictable. Enedis, the French company in charge of national electricity distribution, has been using AI for many years to optimize electricity distribution as well as perform predictive maintenance tasks. Using AI helps them detect in advance which equipment is likely to fail and needs to be taken care of before it happens.

AI will also enable tremendous progress in agriculture by enabling data-driven farming and improving pest detection and crop health monitoring. For years we've been collecting data about a vast number of things, but we lacked tools to help find patterns in that data. What GenAI recently enabled is the ability to digest those results, recognize patterns, compare it with existing industry data, and offer practical improvements. According to McKinsey & Company, GenAI can bring up to $250B of new value in the farming industry.[2] We are all familiar with the examples where AI is helping doctors identify breast cancers, and recently Harvard Medical School mentioned they are working on a ChatGPT-like AI platform to perform a wide range of cancer evaluation tasks.[3] The same approach is true for analysis of satellite images to better predict future natural catastrophes and prevent disasters from happening. As I am typing these lines, Los Angeles fire crews are still battling with one of the biggest fire tragedies of our time. Several entrepreneurs have started companies, such as EXCI in Australia, that use AI to help detect fires early and provide instant notification.

Finally AI tools will permit us to focus on the largest issues to fix and bring a holistic viewpoint on the solutions. AI will be a contributor in mitigating the impact of climate change: weather forecasting, operational scheduling, battery storage optimization, resources optimization, and climate modeling.

Personal Sustainability

Finally, I cannot talk about sustainability and AI without including the topic of personal sustainability. Personal sustainability is about understanding how each of us can find strategies and tools to manage our time and our energy while on a path to fulfill our lives. Every individual is unique and the thousands of studies on mental health, the ideal diet, time management, etc., agree on one thing: there is no single magical recipe that works for everyone.

When defining health, career, and legacy goals, we often lack an accountability partner that knows us perfectly and gives instant feedback, analyzes if a situation is good for us or not, and provides the right tools and techniques to help us progress with our goals. This is the reason why I decided to use the GPT Builder feature offered by ChatGPT to create a doppelganger of myself. I fed this custom-built doppelganger with the obvious documents: my resume, my achievements, my Linked profile, as well my writings. However, I decided to push the concept further and uploaded personality tests I have been taking over the year and the notes I have been taking about myself, my aspirations, my goals, and my failures. I trained the doppelganger by having several chats with it on topics such as what I like and what I don't like. I validated some answers, reworked others, and had it reflect on situations that I lived in the past. I went as far as sharing personal email conversations that didn't result in my expected outcomes.

This doppelganger is now my true coach, supporting me to sort out new opportunities, outline potential issues linked to projects, and it goes as far as recommending to me a schedule to balance my time and energy to achieve my goals. As I am writing this chapter, I am planning the next year's activities and defining my objectives with my doppelganger, which has recommended me specific digital and non-digital tools and techniques such as:

- Using a digital tool like Notion for Project Management or Habitica to gamify my habit-building as well as RescueTime to understand how I am spending my time every day.

- Using Midjourney or DALL-E to create inspirational visuals and ensure I keep the playfulness in my activities

- Using a Bullet Journal to track my monthly personal and professional goals

- Journaling five minutes every day for gratitude and reflection. It helps focusing on what's going well and avoid being too perfectionist

With the help of my development coach, I implemented strategies leveraging my doppelganger to reach flow state and ensure that my relationship to time, money, and energy was sustainable.

I have fed my doppelganger my priorities and big goals for the next 12 months and five years, analyzed and defined my key unique selling abilities, identified what I am very passionate about, and added my biggest roadblocks. Also, I included key questions for it to answer for every activity I get involved in: "How is it linked to any of my goals?", "Can I delegate this activity?", and "Is this activity aligned with my aspirations and values?"

Every day and every week I plan and review my activities to ensure I keep a balance between doing and relaxing. Of course, it requires a certain discipline of sitting down with your GPT; I found it easier to "talk" to it rather than type everything, and it makes the interaction closer to a phone discussion with a "human" coach with the power of automation.

I won't lie, it is very tempting to follow its advice blind, but that wouldn't be sustainable, pun intended. It is important when using an AI-powered tool to always think about how it augments your capabilities and how you remain the sole decision maker around any advice you want to follow. At the end of each week, I celebrate the wins and ask my doppelganger to generate an image and create a virtual book of the wins of the year.

As human beings, we all possess an innate ability to express our emotions and experiences creatively. This unique ability makes us all artists at heart, whether we channel our feelings with words, colors, music, or any other medium. The essence of our creativity is not just the end product but lies in the process of translating our deepest feelings and inner words into something tangible that can speak to others.

I strongly believe that AI-powered tools like Open AI Sora or tools that generate music will democratize art, making it accessible for everyone to express their feelings and stories. Sora and similar innovations offer us the keys to unlock our artistic doors, provided that we are willing to learn and embrace these tools, rather than remain passive spectators. It is enabling us to live more purposeful and more sustainable lives.

AI, like any technology, is really about what we choose to make of it. The integration of AI in my work and life helped me discover that scaling is all about discipline rather than efficiency and quantity. AI can amplify both our strengths and weaknesses, magnify clarity or noise. To navigate this, we must remain grounded on what truly matters: ourselves, our fellow human beings, and our planet. Sustainability is not about mitigating harm; it is about designing systems that allow life to flourish—ours and the Earth's.

AI is ultimately a mirror of our intentions. If we approach it with purpose and accountability, it can help us design a life of more connection, creativity, and resilience. After all, the goal is not to survive in an AI-driven world but to flourish in harmony with the planet and our fellow human beings.

Chapter Endnotes

1. https://newsletter.pessimistsarchive.org/p/telegraph-doomers-of-the-19th-century?utm_source=publication-search

2. https://www.mckinsey.com/industries/agriculture/our-insights/from-bytes-to-bushels-how-gen-ai-can-shape-the-future-of-agriculture

3. https://news.harvard.edu/gazette/story/2024/09/new-ai-tool-can-diagnose-cancer-guide-treatment-predict-patient-survival/

About the Author

Arnaud Blandin is a visionary leader and expert in sustainability, impact measurement, and ESG frameworks. He helps board members, executives, and tech entrepreneurs embed sustainability in their growth strategies. Leveraging his global experience and insights from Silicon Valley and Asia, he aligns technology, innovation, and purpose to drive meaningful impacts.

As the founder of The Beyond Institute, Arnaud has pioneered transformative methodologies to quantify and amplify the environmental and social impacts of human activities. With a robust engineering background—including software contributions to NASA's Mars Rover mission—and over 20 years of global experience in business strategy, technology, and sustainable innovation, Arnaud integrates purpose-driven leadership into the core strategies of Fortune 500 companies, startups, and private equity firms.

Arnaud has partnered with organizations such as ENGIE, Michelin, Johnson & Johnson, Oracle, and Temasek, driving impactful projects like circular economy initiatives in the tire industry, AI-based sustainability strategies for aquaculture, and the creation of data-driven platforms for global energy companies. As a sought-after educator and mentor, he guides over 500 executives annually through programs at INSEAD, HEC Paris, CEDEP, and SKEMA, among others.

An advocate for holistic transitions, Arnaud's work spans food systems, well-being, biodiversity, and the future of capitalism. His thought leadership has shaped frameworks for global challenges such as climate change, resource optimization, and social equity. A prolific contributor to platforms like Creative Destruction Lab and the Monaco Ocean Challenge, he remains deeply committed to nurturing the next generation of sustainable leaders.

As an entrepreneur, Arnaud has founded and exited multiple ventures across sectors including digital technology, management consulting, and wines and spirits. Beyond his professional endeavors, he continues to inspire by connecting cultures, embracing innovation, and advocating for purposeful living. Whether mentoring young minds or collaborating with global leaders, Arnaud's mission is clear: turn sustainability into a catalyst for growth and meaningful change.

Email: arnaud@arnaudblandin.com
Website: www.arnaudblandin.com

CHAPTER 5

BECOMING AI LITERATE

By Sean Musch
Founder and CEO, AI & Partners; Risk
Amsterdam, Netherlands

By Michael Borrelli
Director, AI & Partners; Regulation
London, England, United Kingdom

> *Education is the most powerful weapon with*
> *which you can change the world.*
>
> —Nelson Mandela

The artificial intelligence (AI) revolution is transforming the world at lightning speed, bringing smart machines, intelligent systems, and groundbreaking innovations into every corner of our lives. But as exciting as this era may be, thriving in it requires more than a passing familiarity with technology—it demands AI literacy. Far from being optional, we believe that AI literacy is becoming a critical skill for navigating the complexities of an AI-driven world. The European Union (EU) AI Act, the world's first comprehensive law on AI, underscores the importance of AI literacy in Article 4, requiring both education and training in AI to ensure its ethical, responsible, and effective use.

We have written this chapter to explore why AI literacy matters, not just for technologists, but for entrepreneurs, educators, leaders, and anyone ready to take charge in an AI future, and how you can become a leader. With the right mindset and strategies, AI literacy can empower us all to move beyond fear of the unknown and embrace AI's transformative potential with confidence and responsibility.

Mindset Shift

In the fast-evolving world of AI, one thing is clear: success requires a mindset shift. AI literacy—essential for navigating the complexities of this new era—is not just about technical know-how; it's about cultivating the right attitudes and perspectives to leverage AI effectively. Developing an AI mindset is the foundation of this shift, and we always begin with three critical steps: (1) understanding and awareness, (2) ethical and responsible use, and (3) innovation and creativity.

AI literacy starts with understanding what AI can and cannot do. It's easy to get swept up in the hype—believing AI is an all-knowing oracle or, conversely, dismissing it as an overhyped trend. Neither perspective is productive. A foundational understanding of AI's capabilities and limitations empowers individuals to make informed decisions and critically assess its role in their lives and industries. Take financial literacy as its financial equivalent. If we do not have the basic knowledge of what money is, how to save, and all these other elements, we cannot interact in the modern-day world. How is AI any different in this sense?

Think about it this way: AI isn't magic; it's a tool. And like any tool, its value depends on how well you understand its strengths and weaknesses. For example, an entrepreneur launching a product might use AI for market analysis. Without understanding AI's analytical potential—or its limitations—they risk missing key insights or misinterpreting data, leading to flawed strategies. To paraphrase Gregory Titelman's idiom, which alludes to a catch or mysterious element hidden in the details, "The devil is in the details." While AI use may appear simple, the details are complicated and likely to cause problems. But with a grasp of AI's mechanics, people can harness it as a strategic ally, turning data into actionable intelligence.

Awareness also fosters critical thinking. Leaders with an AI mindset ask the right questions: *What problems can AI solve? Where does it fall short? How does it complement human expertise?* Being curious is the very essence of being human, although the rise of autonomous forms of AI (referred to as "agentic AI") can challenge this view. These questions open the door for thoughtful integration of AI into workflows, ensuring it adds value rather than creating unnecessary complexity.

"With great power comes great responsibility." This proverb, popularized by Spider-Man in Marvel comics, films, and related media, is particularly true for AI. As AI systems grow more influential, the ethical stakes become higher. AI literacy demands an understanding of these ethical implications, shaping a mindset that prioritizes transparency, fairness, and accountability.

Bias is one of the most significant challenges in AI. Algorithms, no matter how sophisticated, reflect the data they're trained on. If that data contains biases, the AI will perpetuate them—sometimes amplifying inequalities in ways we may not even notice. Leaders with an ethical mindset are proactive in identifying these risks, ensuring their AI systems serve diverse audiences equitably. If we don't address bias at the input stage of an AI, how can we expect equitable outcomes?

As always, we draw inspiration from the EU AI Act. Take the example of a recruitment platform powered by AI likely covered under Annex III of the EU AI Act. If the underlying algorithm favors certain demographics due to historical data biases, it could reinforce discrimination rather than create opportunities. If leaders understand these risks, they can implement safeguards like regular audits and bias checks, fostering a fairer and more inclusive system.

Ethical AI isn't just about avoiding harm; it's about building trust. Transparent decision-making processes—where users understand how and why an AI reached a conclusion—enhance credibility. Likewise, compliance with privacy regulations and robust data protection practices demonstrate respect for users' rights, aligning AI systems with fundamental rights, such as those referred to in Recital 1 of the EU AI Act. Funnily enough, being biased has become an inherent trait—we all have biases. For AI, judgement of an outcome driven by bias depends on the context in which it was applied.

An AI mindset doesn't stop at understanding and ethics—it thrives on innovation. AI literacy unlocks creativity by revealing new ways to solve problems and improve processes. Rather than viewing AI as a replacement for human ingenuity, those with an innovative mindset see it as a collaborator, amplifying what humans do best: taking action. Alfred North Whitehead would say, "Ideas won't keep. Something must be done about them."

Consider the healthcare industry, where AI literacy has potential to help clinicians develop predictive models that anticipate patient needs. These innovations not only can improve patient outcomes but also reduce costs, changing how care is delivered. This kind of breakthrough stems from a mindset that sees AI as a partner in problem-solving, not a competitor. This echoes the thoughts of leaders, such as Silvio Savarese, executive vice president and chief scientist at Salesforce AI Research, who recognises AI's potential as a "superpowered collaborator".

The same principle applies across industries. Manufacturing leaders, like Amazon, are using AI to optimize supply chains, educators are personalizing learning experiences, and artists are pushing creative boundaries. The common thread? A willingness to embrace AI's potential while remaining grounded in its realities.

Collaboration is another hallmark of the innovative AI mindset. A synergy of diverse perspectives—technologists, designers, ethicists, and end-users—helps organizations to uncover unique applications for AI. This cross-disciplinary approach ensures that AI solutions are not only effective but also inclusive and human-centric—a core facet of the EU AI Act.

We finish this part by noting how embracing an AI mindset is not an overnight transformation—it's a journey. It starts with education, whether through formal training, hands-on experience, or simply staying curious about AI's evolving role in society. The EU AI Act recognizes this by emphasizing the need for AI literacy under Article 4, urging stakeholders to equip themselves with the knowledge and skills to use AI responsibly. To be frank, this mindset shift is about more than personal growth; it's about creating a future where AI enhances, rather than detracts from, our shared humanity.

Becoming Strategic

Navigating the world of AI without a strategy is like trying to win a game of chess without understanding the rules—you're likely to be outmaneuvered. The good news? Building effective AI strategies is entirely within reach, even if the technology feels overwhelming at first. Think of it as a journey: the more you learn, the more confident you become in using AI to your advantage. Ralph Waldo Emerson supports this view with his famous quote, "Life is a journey, not a destination."

Let's start with the basics: understanding what AI can and cannot do. This isn't about becoming a computer scientist overnight. It's about grasping the essentials—AI's strengths, its limitations, and how it fits into your world. We see a growing consensus that you have to be very technical to embrace AI, which simply isn't true—everyone has a role to play.

Take a practical example. Imagine you're leading a team tasked with implementing a customer service chatbot. Without understanding AI's nuances, you might end up with a chatbot that answers "How can I help you?" to every query, frustrating your customers more than helping them. But with AI literacy, you'd know how to design a system that hands off complex issues to human agents, striking the perfect balance between automation and personalization.

This level of understanding ensures that your AI initiatives align with your goals, resources are used wisely, and risks are managed proactively. It's like having a GPS for your AI journey—helping you make informed decisions and avoid costly wrong turns.

Now, let's add a layer of heart to your strategy. AI isn't just about efficiency or innovation; it's about people. To use AI responsibly, we need strategies rooted in fairness, inclusivity, and transparency. As most management theories go, "people" remains a core element of the PPT framework for businesses: people, process, and technology—we are still needed!

Think of it this way: every AI decision you make ripples outward, affecting customers, employees, and society at large. A biased hiring algorithm, for example, could unintentionally favor certain groups over others, tarnishing your reputation and limiting diversity in your workforce. Take inspiration from the Dalai Lama, when he noted, "Just as ripples spread out when a single pebble is dropped into water, the actions of individuals can have far-reaching effects."

The solution? Build guardrails into your strategy. Establish codes of practice that champion fairness, involve diverse perspectives in AI development, and prioritize transparency. Article 56 of the EU AI Act, for instance, is an excellent guide—it emphasizes ethical principles like minimizing bias and ensuring accountability for general-purpose AI models, like ChatGPT. When you embed ethics into your strategy, you're not just avoiding legal trouble—you're building trust. And trust is the bedrock of long-term success in the AI era.

Here's where things get tricky. Regulations, like those in the EU AI Act, can feel like roadblocks to innovation—some even use the term "business prevention". But what if they're actually stepping stones? Take regulatory sandboxes, for example. These "safe zones" allow companies to experiment with AI within clearly defined boundaries, balancing creativity with compliance. It's like practicing a new recipe in a test kitchen before serving it at a big dinner party.

Another game-changing strategy? Investing in general-purpose AI models—systems designed to adapt across various applications while meeting ethical and regulatory standards. OpenAI, for instance, has mastered this approach, creating versatile tools that both push the boundaries of what's possible and respect the rules of the game.

The key takeaway? Compliance doesn't have to stifle innovation. With the right strategies, it can drive you to think smarter and create better. But, it requires an upfront investment of both time and energy to gain understanding.

Finally, let's talk about keeping up. AI isn't a "set it and forget it" tool. It's a living, evolving technology, and staying ahead means committing to continuous learning.

Start by training yourself and your teams. This doesn't mean overwhelming them with technical jargon but equipping them with the practical skills needed to use AI effectively. Encourage collaboration between developers, users, and regulators—it's amazing what fresh perspectives can uncover.

And don't forget iterative improvement. Companies like Amazon have mastered this with their logistics systems, constantly tweaking algorithms to achieve peak performance. The result? Happier customers, smoother operations, and a competitive edge that's hard to beat.

Becoming strategic with AI isn't about mastering every technical detail—it's about approaching AI with curiosity, responsibility, and a clear vision. It's about using AI as a tool to enhance understanding, promote ethics, spark innovation, and foster growth.

So, the next time you feel daunted by AI's complexities, remember: strategy is your compass. With it, you're not just surviving the AI revolution—you're thriving in it.

Thrive, Not Survive

The AI revolution is here, and it's not waiting for anyone to catch up. The question isn't whether you'll adapt but how you'll thrive in this new landscape. The good news? Thriving is entirely possible—if you embrace the right mindsets and strategies.

At its core, thriving in an AI-driven world starts with education. AI literacy isn't about memorizing technical jargon or becoming a coder overnight; it's about understanding what AI can do, where it falls short, and how to use it to your advantage.

Imagine standing at the base of a mountain, unsure how to climb it. It feels daunting, right? Without the right tools and knowledge, the journey would feel overwhelming—maybe even impossible. But with a little preparation, some guidance, and a dose of courage, you'll find yourself scaling that peak. AI literacy is your preparation, your guide, and your safety rope all in one.

When you understand AI's capabilities and limitations, you're better equipped to align it with your goals, make informed decisions, and avoid costly missteps. Whether you're an entrepreneur brainstorming your next business idea, a leader navigating organizational change, or simply a curious bystander looking to learn more about how AI will affect you, AI literacy empowers you to act with clarity and confidence.

But knowledge alone isn't enough. Thriving in the AI era requires a mindset shift—one that's open to innovation, willing to embrace uncertainty, and focused on growth. Let's face it: AI can be intimidating, and we've found this ourselves. It's easy to feel overwhelmed or worry about being left behind. But the most successful people in this space don't let fear dictate their actions. Instead, they view AI as an opportunity to expand their horizons, solve complex problems, and unlock new possibilities.

Think of emerging entrepreneurs who can build thriving businesses by leveraging AI to personalize customer experiences or automate tedious tasks. They didn't start with all the answers, but they had the curiosity and determination to explore what AI could do for them. They saw challenges not as obstacles but as stepping stones.

And that's the key: thriving with AI isn't about perfection. It's about persistence. It's about taking small, steady steps toward a future where AI enhances—not replaces—your unique skills and talents.

Of course, a positive mindset needs to be backed by smart strategies. Without a plan, even the most optimistic outlook can falter.

So how do you turn your AI aspirations into actionable outcomes? Start by aligning AI initiatives with your goals. Ask yourself: "How can AI make my work easier, faster, or more impactful?" Whether it's streamlining operations, enhancing customer engagement, or driving innovation, clear objectives help you focus your efforts.

Next, prioritize ethics and inclusivity. AI isn't just about efficiency; it's about doing what's right. Strategies rooted in fairness, transparency, and diversity aren't just good for society—they're good for business. When people trust that your AI systems are designed with their best interests in mind, they're more likely to support and adopt them.

Finally, commit to continuous learning. AI is evolving at breakneck speed, and staying ahead requires adaptability. Invest in ongoing training, seek out diverse perspectives, and never stop asking questions. Remember: every challenge is an opportunity to grow.

The beauty of AI is that it's not a solo journey. Whether you're an educator, a corporate leader, or a budding entrepreneur, collaboration is your secret weapon. Sharing knowledge, learning from others, and working together to solve problems, means we can shape an AI future that benefits everyone.

As you navigate this exciting new world, keep this in mind: AI is a tool, not a threat. It's here to amplify your potential, not diminish it. The key to thriving isn't about having all the answers—it's about staying curious, taking action, and building a future you can be proud of.

So, let's embrace this moment together. We can all educate ourselves, challenge our fears, and create strategies that not only help us survive but allow us to thrive in the AI revolution. Because the future isn't written yet—it's waiting for us to shape it.

About the Authors

Sean Musch, CEO and founder of AI & Partners, has an extensive background in the entertainment industry, specifically film and art, and has a specialization in design. Alongside this, Sean has more than a decade of experience in the professional services sector, including holding the position of a tech accountant for five years. Sean knows about auditing and has helped with an IPO on the New York stock exchange. As well as being a compliance expert, he has deep expertise in implementation aspects of audit and assurance engagements, and has been working with the largest global tech MNEs over the past five years.

Michael Charles Borrelli, director of AI & Partners, is a highly experienced financial services professional with over ten years of experience. He has held executive positions in compliance, regulation, management consulting, and operations for institutional financial services firms, and has consulted for FCA-regulated firms on strategic planning, regulatory compliance, and operational efficiency. In 2020, Michael set up the operations model and infrastructure for a crypto-asset exchange provider and has been actively engaged in the Web 3.0 and AI communities over the last four years. He currently advises a host of AI, Web3, DLT, and FinTech companies.

Email: contact@ai-and-partners.com
Website: https://www.ai-and-partners.com/

CHAPTER 6

AI AS A GIFT: TRANSFORMING HEALTHCARE, LIFE, AND DREAMS

By Jaspreet Chager, BSc. Phm
Single Mother, Healthcare Leader, AI Literacy Advocate
Toronto, Ontario, Canada

Sometimes the smallest step in the right direction can be the biggest step of your life. Tiptoe if you must, but take the step.

—Naeem Callaway

I used to think leadership was about having all the answers. After nearly 20 years driving healthcare transformation—implementing pharmacy services, shaping policy at the highest levels, serving on boards and external committees—I was confident in my ability to solve any challenge. Give me a complex system to optimize, a team to lead, or a strategic initiative to implement, and I would find a way forward. My professional toolkit seemed complete.

Life, however, has a way of humbling even the most seasoned leaders. When I left my marriage, I discovered that all my expertise in healthcare leadership couldn't prepare me for the profound solitude of those quiet evenings after my children went to sleep. During my parenting time, every decision, every responsibility rested solely on my shoulders. My world narrowed to its essential core: my children, my parents, and the weight of knowing I had to somehow keep everything afloat—from managing finances I'd never handled before to maintaining my professional trajectory—all while ensuring my children had the support and stability they needed.

That's when AI entered my life like an unexpected gift. It wasn't through some grand strategic initiative or corporate implementation. Instead, one late night, after tucking my children into bed and facing the mountain of tasks ahead, I discovered an AI writing tool that could help draft the documents and emails I'd been tackling in those quiet hours. This simple discovery became my first glimpse of possibility—a way to handle necessary tasks more efficiently, so I could focus on what truly mattered.

But AI gave me something far more valuable than just time—it gave me the space to dream again. As the technology helped me streamline various aspects of my life, I found myself with enough mental bandwidth to envision possibilities I'd never dared to consider. I began to imagine a life beyond just managing, beyond just balancing. I could see a future where I might help others understand and harness this technology to transform their own lives.

Now, let me be clear—AI didn't solve everything. It couldn't cut the grass, take out the garbage, or handle parent-teacher conferences. Like any powerful tool, AI comes with its share of challenges and potential pitfalls. There are valid concerns about privacy, bias, and the risk of over-reliance on technology. But I've learned that the best way to address these challenges isn't to avoid AI—it's to understand it. AI literacy becomes our pathway to using these tools thoughtfully and ethically.

In the rest of this chapter, I'll explore how AI literacy can transform four key domains—healthcare, women's empowerment, small business success, and professional development. While these might seem like separate spheres, they converge in one fundamental truth: when we understand how to use technology thoughtfully and ethically, it frees us to focus on what truly matters. Even in our most overwhelming moments, AI literacy can help us find the breathing room to not just survive, but to imagine and create the life we want.

Healthcare Empowerment: Moving Beyond Data Overload—The Promise and the Pitfalls

Healthcare loves its technology—we're swimming in electronic health records, automated billing systems, and endless documentation requirements. Yet ironically, a 2024 survey reports that nearly half of clinicians cite "inefficient digital processes" as a core driver of burnout. Let me paint you a picture: imagine highly trained professionals spending hours clicking through clunky interfaces instead of caring for patients. It's like buying a Ferrari and using it only to drive to the mailbox—we've somehow turned powerful tools into elaborate obstacles.

Having overseen COVID-19 vaccine rollouts across hundreds of locations, I witnessed firsthand how administrative burden pushed healthcare workers to their breaking point. The pandemic didn't just challenge us medically—it exposed and exacerbated the systemic issues that were already driving professionals from the field. Now we're facing a critical shortage of healthcare workers, with many leaving the profession entirely. I've watched brilliant providers arrive early and leave late, not because of complex patient cases, but because they're trapped in what I call the "digital maze"—endless screens, passwords, and forms that seem designed by someone who's never spent a day in healthcare. And while AI can't solve everything (it can't shovel snow from the hospital parking lot—yet), it can tackle the administrative burden that's crushing the spirit out of healthcare.

The numbers tell a sobering story: nurses spend up to 30% of their shift just entering data. That's not what they went to school for, and it's certainly not what draws passionate people to healthcare. During the vaccine rollouts, I watched our already-strained workforce be pushed to the limit—not just by the demands of patient care, but by the crushing weight of documentation, inventory management, and regulatory requirements. One in five healthcare workers have quit since 2020, and it's not hard to understand why. It wasn't the patient care that exhausted them—it was the endless administrative tasks, the constant switching between systems, and the mind-numbing data entry that left them too drained to bring their best selves to patient interactions.

Then came AI, offering a different way forward. Pilot studies show AI-based intake can slash patient check-in times by 35% while smart scheduling tools cut staff overtime by 15%. Radiology departments using AI for image pre-screening have improved early detection rates by 7% to 8%. These aren't small wins—they represent precious minutes and hours given back to healthcare providers at a time when we're facing critical staffing shortages across the industry.

This rebalancing—letting AI handle the mechanical while humans focus on the humane—could be our path forward during a critical healthcare staffing crisis. With one in five healthcare workers having left the profession since 2020, and many more considering leaving, we can't afford to ignore solutions. The pandemic didn't just challenge us medically—it exposed and exacerbated the systemic issues driving professionals from the field. Studies consistently link lower administrative burdens to higher job satisfaction and better patient outcomes. However, success depends entirely on thoughtful implementation. Healthcare organizations must invest in proper training, ensure robust data protection protocols, and maintain clear guidelines about when human judgment must prevail over automated processes.

The premise is simple though the implementation requires care—empowerment happens when we let technology handle what it does best: freeing healthcare professionals to do what only humans can—connect, empathize, and innovate. But this isn't about blind adoption. It's about developing the AI literacy to understand both the potential and the limitations of these tools. In an era where healthcare worker burnout has reached crisis levels, AI offers more than just efficiency—it offers hope for rebuilding a sustainable, human-centered healthcare system. And maybe, just maybe, it can help our healthcare heroes get home in time for dinner with their families. Speaking of dinner time—that's where another dimension of this story unfolds.

Women's Empowerment: Breaking Free from Invisible Labor

The burden of administrative overload in healthcare mirrors a broader challenge that women face across all sectors—the weight of invisible labor. After my children went to bed, the house would fall into a particular kind of silence that only single parents understand. It wasn't just quiet—it was the weight of knowing every undone task, every decision, and every next day's challenge rested solely on my shoulders. During my parenting time, there was no backup, no one to share the mental load of running a household while maintaining a demanding career. My support system had narrowed to my parents, which was absolutely enough, but it meant I had to find new ways to manage everything else.

Research consistently shows that women perform a disproportionate amount of unpaid labor—both at work and at home. We're often the ones managing the invisible tasks that keep both households and workplaces running smoothly, yet rarely do we get recognized for this additional mental load. Ironically, a 2023 workforce analysis reveals that women adopt AI tools at a 7% to 12% lower rate than men, often because they're too busy managing everything else to explore new technology or because they doubt their ability to master it.

This hesitation creates a challenging cycle—by not embracing AI tools, many women remain trapped in routine tasks that prevent them from pursuing leadership opportunities or imagining different possibilities. Yet the potential for transformation is remarkable. When women harness AI effectively, they can break free from these historical constraints. That extra hour not spent on administrative tasks becomes an hour for strategic thinking, professional development, or pursuing long-held dreams.

Of course, we must acknowledge the challenges. Workplaces need to ensure women aren't left behind in AI training opportunities. We must be vigilant about biases embedded in AI systems that might perpetuate gender stereotypes. But for those who embrace it, AI can be a powerful equalizer, giving women back the hours—and the energy—they need to shape not just their own futures but entire industries.

The impact extends beyond individual efficiency. In a recent global survey, women entrepreneurs using AI tools reported not just time savings, but increased confidence in their decision-making. They're leveraging AI for everything from market analysis to financial planning, tasks that once required expensive consultants or extensive training. The technology is democratizing access to expertise in a way that particularly benefits women who might not have traditional business networks or resources.

Think about the implications: a single mother can run sophisticated business analyses after her kids are in bed. A female healthcare provider can automate her documentation and spend more time advancing her research. A woman returning to the workforce can quickly update her skills using AI-powered learning tools. These aren't just convenience features—they're game-changers for women who've historically had to choose between career advancement and life's other demands. This potential for transformation extends beyond individual empowerment to reshape entire organizations, particularly those operating on a smaller scale.

SMEs: The Heart of Healthcare Innovation

Healthcare isn't just about large hospital systems. A significant portion of health services operate as small and medium-sized enterprises (SMEs)—clinics, specialized pharmacies, outpatient labs—businesses with fewer than 250 employees, often grappling with the same resource constraints as any other small organization. While my experience is in healthcare, these challenges resonate across all SMEs, regardless of industry. Whether you're running a small medical practice, a local retail business, or a consulting firm, the fundamentals remain the same: limited resources, multiple roles per person, and the constant juggle of operational and strategic demands.

AI can be a lifeline for these smaller organizations. A 2024 report found that SMEs incorporating AI technologies saw overhead shrink by up to 20%. For healthcare providers, this might mean using AI tools to forecast inventory needs, avoiding overstock or shortages

of key medications. In other industries, it could mean automating customer service inquiries or streamlining documentation processes. The principle remains consistent across sectors: when routine tasks are automated, staff can focus on higher-value activities that drive business growth and improve customer service.

Still, SMEs often approach AI adoption cautiously, concerned about costs, data privacy, and the learning curve. This is particularly crucial in healthcare, where patient data security is paramount. The key isn't implementing every available technology—it's identifying specific challenges that consume the most resources and finding appropriate AI solutions. Whether that's streamlining documentation, managing inventory, or handling customer communications, the goal remains consistent: freeing human talent for tasks that require judgment, creativity, and personal connection.

What makes this particularly exciting is how AI democratizes access to sophisticated capabilities. Small practices can now leverage tools once available only to large corporations, without extensive IT departments or massive budgets. A local business may be able to implement inventory systems rivaling major chains, while maintaining the personal touch that distinguishes smaller organizations from their larger competitors.

This democratization, however, comes with responsibility. Organizations must approach AI adoption with careful consideration of privacy, security, and ethical implications. Success depends not just on implementing the technology, but on ensuring staff have the knowledge and confidence to use it effectively. And that brings us to perhaps the most crucial element of this transformation: the importance of proper training and education in AI tools. Without it, even the most powerful technology remains underutilized or, worse, misused.

Upskilling: Securing Your Future in a Tech-Driven World

A 2025 projection suggests that roles requiring AI literacy will soon dominate many industries. This isn't just about staying competitive—it's about using these tools responsibly and effectively. For clinicians, that might mean learning to interpret AI-driven diagnostic suggestions while maintaining critical human judgment. For SME owners, it could be adopting digital marketing tactics while ensuring customer data privacy.

In my own journey, the decision to systematically learn AI tools marked a clear turning point. But this wasn't about blind adoption. Every step required careful consideration: which tools were trustworthy? How could we verify outputs? What tasks should remain firmly in human hands? This measured approach to upskilling—understanding both the potential and limitations of AI—became fundamental to using it effectively.

The mindset shift is crucial. While some fear AI might make their roles obsolete, the reality is more nuanced. The goal isn't to replace human judgment but to enhance it. The real breakthrough comes when people realize how AI tools, used thoughtfully, can give them back time for meaningful work while maintaining the quality and integrity of their profession.

Weaving It All Together: Resilience Through Empowerment

When I look back at my transformation—from those quiet nights of uncertainty to building a business focused on AI literacy and education—I see more than just my own story. I see a blueprint for possibility. Healthcare professionals can rediscover their passion for healing when freed from administrative burdens. Women can reclaim hours once lost to invisible labor. Small businesses can compete with industry giants. The common thread? Understanding how to harness AI thoughtfully and effectively.

But this isn't a story about technology alone—it's about empowerment through understanding. AI literacy becomes our compass, helping us navigate both the opportunities and challenges these tools present. Yes, there are valid concerns about privacy, bias, and the risk of over-reliance on technology. That's precisely why education and thoughtful implementation matter so much.

The future I envision isn't about replacing human judgment—it's about the potential to enhance our uniquely human capabilities. AI can be a gift, but only if we learn to use it wisely. When we do, we create space for what truly matters: deeper patient connections, strategic thinking, innovation, and perhaps most importantly, the mental bandwidth to dream bigger than we ever thought possible.

About the Author

Jaspreet Chager is a single mother, pharmacist, healthcare leader, and entrepreneur who discovered how AI could unlock a life she never imagined possible. With over 20 years of experience in Canadian healthcare, including C-suite and board positions, she experienced firsthand how AI could transform not just careers, but entire lives. As founder of AI Simplify Now, she empowers professionals, particularly women, across healthcare and other industries to envision and achieve possibilities through AI literacy and education. Her approach combines executive-level strategic thinking with practical solutions learned from navigating life's complexities. Jaspreet is passionate about ensuring AI literacy reaches those who need it most, especially women and underrepresented groups, believing that technology should serve as an equalizer, helping everyone create their own version of an extraordinary life.

Email: jchager@aisimplifynow.io
Website: www.Chager.org

CHAPTER 7

AI REVOLUTION IN HEALTHCARE: DRIVING VALUE THROUGH ADVANCED ANALYTICS AND INSIGHT

By Victor Collins
Associate Director Data & AI at Impact Advisors, LLC
Hillsborough, North Carolina

AI is perhaps the most transformational technology of our time, and healthcare is perhaps AI's most pressing application.
—Satya Nadella, CEO, Microsoft

For decades, healthcare organizations have pursued digital transformation to improve operations, care delivery, and administrative efficiency. However, what many organizations have achieved is not true transformation but rather digital reformation. Instead of fundamentally reimagining workflows, they digitized legacy processes, reinforcing existing structures and inefficiencies.

The implementation of electronic health records (EHRs) exemplifies this challenge. EHRs digitized vast amounts of clinical, financial, and operational data, but many healthcare systems simply replicated their paper-based workflows in a digital format. While automation and data integration brought some improvements, these efforts did not fully transform how care is delivered or how decisions are made. Silos remained intact, limiting data flow, collaboration, and innovation across departments. The healthcare industry, long beleaguered by rising costs, operational inefficiencies, and the challenge of delivering personalized care at scale, finds itself at a critical inflection point.

Now, with the rise of artificial intelligence (AI), healthcare organizations have a second chance—a chance not just to automate existing tasks but to redefine care delivery, operational models, and patient engagement. This is the dawn of an intelligence revolution, one that challenges every assumption about how we structure healthcare systems, measure outcomes, and achieve success.

The Digital Misstep—Learning from Past Efforts

The promise of digital transformation was to revolutionize the healthcare industry by connecting data, systems, and processes. Despite spending billions of dollars and decades of effort, today many digital clinical innovations have either failed outright or have been met with hindrance and resistance by health professionals—often visions of AI in action which failed to function in practice. Researchers have found that the primary cause of this is the lack of consideration for human factors, including providers' needs, values, and ways of working.

For many healthcare providers, digital transformation fell short. EHRs, billing systems, and resource management platforms became systems of record that operated within departmental silos. Each department—patient services, finance, clinical operations, and IT—optimized its own workflows without sufficient integration across the organization. These tools reformed operations but did not transform them. Instead of breaking down silos, they fortified them, making systemic collaboration more difficult.

The Intelligence Revolution: AI as a Game-Changer

The rise of AI presents healthcare with a unique opportunity to do things differently. Unlike digital technologies that focus primarily on automation, AI introduces the capacity to learn, adapt, and provide insights that can drive continuous improvement. AI can enable organizations to go beyond simply doing the same tasks faster; it invites them to rethink *which* tasks should be done, *how* they should be performed, and *why* they are important.

AI offers the potential to create AI-first healthcare organizations, where intelligence is embedded across every function—from diagnostics and treatment planning to operational efficiency and patient engagement. This is about more than automation; it is about innovation and reinvention. The question for healthcare leaders is clear: will AI be used to accelerate outdated processes, or will it be leveraged to fundamentally transform care delivery and patient outcomes? In the words of entrepreneur Mark Cuban, "There are only two types of companies in this world: those who are great at AI and everybody else. If you do not know AI, you are going to fail, period, end of story. You must understand it, because it will have a significant impact on every single thing that you do. There's no avoiding it."

AI's Transformative Impact on Healthcare

Healthcare organizations are increasingly adopting advanced AI technologies to solve complex problems. Each of these innovations—generative AI, ambient AI, agentic AI, and intelligent automation—plays a crucial role in improving both clinical and operational workflows.

Generative AI in Healthcare

Generative AI can generate new data, insights, and content, helping healthcare professionals streamline operations and enhance patient care. For example, large language models (LLMs) can automate clinical documentation, reducing the administrative burden on physicians and freeing time for direct patient interaction. GenAI generates new content, including clinical notes, patient communication, and synthetic training data. Its capacity to analyze massive datasets helps healthcare providers streamline operations and improve patient outcomes.

- Use case: a hospital uses generative AI to create automated discharge summaries and personalized follow-up instructions, improving both clinician efficiency and patient understanding.

- Impact: generative AI accelerates research by synthesizing insights from medical literature and clinical trials, driving advancements in precision medicine.

- Challenges: without proper validation, AI-generated content can be inaccurate, potentially compromising patient safety.

Ambient AI in Real-Time Care Monitoring

Ambient AI continuously monitors patient conditions, gathering data without direct human input. This is particularly valuable in high-risk environments such as intensive care units (ICUs) and emergency departments, where timely interventions can save lives.

- Use case: an ICU deploys ambient AI to monitor patients' vital signs, detecting subtle changes that signal early sepsis and prompting rapid intervention.

- Impact: early detection reduces mortality rates and prevents critical events.

- Challenges: excessive alerts can lead to "alert fatigue" among providers and caregiver staff, requiring careful system configuration to prioritize critical warnings.

Agentic AI for Workflow Optimization

Agentic AI goes beyond passive data analysis by autonomously managing workflows and initiating actions based on real-time information. It can optimize resource allocation, streamline scheduling, and coordinate care delivery. Agentic AI autonomously manages workflows and resources, dynamically optimizing operations in real time.

- Use case: during a patient surge, an emergency department's agentic AI reallocates staff and equipment to prioritize urgent cases.

- Impact: agentic AI improves operational efficiency, reducing wait times and resource waste.

- Challenges: over-reliance on automation may limit human oversight, posing risks in complex clinical scenarios.

Intelligent Automation for Process Efficiency

Intelligent automation integrates AI with robotic process automation (RPA) to streamline repetitive tasks such as patient intake, claims processing, and billing. Intelligent automation has the potential to transform healthcare operations by enhancing efficiency, reducing costs, and improving accuracy. However, successful implementation requires careful planning, robust governance, and a commitment to workforce engagement. By addressing these challenges, healthcare organizations can unlock the full potential of automation to drive sustainable improvements in care delivery and financial performance.

- Use case: an organization implements an intelligent automation system to handle claims processing. The system automatically extracts data from EHRs, verifies patient eligibility, cross-checks insurance policies, and submits claims. It flags any discrepancies for human review and learns from historical patterns to improve accuracy over time.

- Impact: this significantly reduces the time required by the revenue cycle staff to process claims, thus reducing costs and allowing healthcare providers to accelerate revenue collection. Staff who were previously burdened with manual tasks can focus on higher-priority activities such as patient engagement and compliance management.

- Challenges: many healthcare organizations struggle with fragmented data systems. Integrating automation tools with multiple platforms (e.g., EHRs, billing systems, insurance portals) can be technically complex and costly. Staff may fear job displacement or struggle to adapt to new workflows. Effective change management and communication are essential to foster acceptance.

Breaking Down Silos: Integrated Intelligence Across Healthcare

AI's potential is constrained when implemented in isolated functions. For AI to deliver transformative value, healthcare organizations must eliminate silos and connect workflows across the entire enterprise. The potential of AI in optimizing care delivery processes is compelling. Predictive analytics—powered by AI such as machine learning algorithms and large language models enabling healthcare organizations to anticipate patient needs—optimizes resource allocation and prevents adverse events. It predicts hospital readmissions using random forest models to managing patient flow in emergency departments through reinforcement learning algorithms. Optimizing a scheduling system is far more impactful when it is integrated with predictive models for patient admissions, staffing, and resource availability. This allows data to flow freely between departments, creating a connected enterprise that leverages AI to drive systemic efficiencies and improved outcomes.

Embracing an AI-First Mindset

Achieving AI-first transformation requires a shift in mindset. Traditional business structures categorize work by function—HR, finance, clinical operations—each with its own system of record. However, modern healthcare demands cross-functional collaboration, where intelligence is shared and integrated across roles and departments.

The Future of AI-Driven Transformation

As AI technologies evolve, healthcare organizations can redefine not just how work is done but what is possible. Large language models are transitioning into "large action models," capable of executing tasks based on insights. AI agents will soon be able to act across integrated workflows, accelerating innovation and unlocking new efficiencies.

AI-Driven Transformation

- Automation: routine tasks such as patient intake and claims processing are fully automated.
- Augmentation: clinicians use AI to enhance their decision-making with real-time analytics and predictive models.
- Innovation: AI enables entirely new capabilities, such as AI-driven clinical trials and personalized care delivery at scale.

Governance and Regulatory Challenges

Healthcare organizations will need to increase their focus on improving data quality and implementing responsible and ethical AI solutions, with strong governance frameworks as the foundation to ensure data accuracy, completeness, and privacy compliance. Healthcare organizations must prioritize governance frameworks to navigate this regulatory uncertainty. This includes creating roles such as chief AI officer to oversee compliance, data ethics, and risk management.

Overcoming Barriers to AI Adoption

Despite the promise of AI, healthcare organizations face several obstacles to full-scale implementation. Common barriers include:

Data Quality and Interoperability

Incomplete or inconsistent data can hinder AI performance, requiring significant investment in data governance and integration. The most powerful algorithm using low-quality data will produce low-quality results. Organizations must take advantage of modern data tools, such as data lakes and data fabrics, to extract the maximum value from data. The modern tools enable storage of vast amounts of source data, which can be used for AI training and analytics, ensuring that AI models are trained on diverse, representative datasets.

Workforce Resistance and Adaptation

Resistance to AI adoption, especially among clinicians, is often driven by concerns about technology fatigue and inaccuracies, exacerbated healthcare disparities, confused lines of responsibility, and increased workloads. EHRs have digitized patient records, but these tools have not necessarily made providers' lives easier. AI solutions should be designed to support patient care as a foundational component of the practice of medicine. Comprehensive training programs can help address these concerns as well as the development of tools that provide transparent, interpretable insights for clinicians and patients.

Cybersecurity Threats

As AI systems collect and analyze sensitive data, they become attractive targets for cybercriminals. Robust security measures are essential to prevent breaches and protect patient privacy.

Cost and Resource Constraints

Smaller healthcare providers may struggle to afford AI infrastructure, creating a potential divide in access to innovation. There is a risk we are going to create all kinds of innovative solutions that are only available for sophisticated health systems.

Patient Engagement and Journey

The integration of AI in healthcare has concentrated on enhancing clinical workflows and operational and financial efficiency. However, patient education regarding AI's role in their care remains limited, which can significantly influence the acceptance and effectiveness of AI-driven healthcare solutions.

The Role of Healthcare Leadership in the AI Revolution: Leading Change in the Age of Intelligence

As the AI revolution reshapes healthcare, a pressing question emerges: how will AI redefine the roles of healthcare leaders? The reality is clear—AI is not here to replace healthcare professionals but to empower them. Those who embrace AI and strategically integrate it into their organizations will drive innovation, enhance efficiency, and improve patient outcomes. Conversely, those who resist change risk falling behind. AI adoption is not just a technological shift—it is a leadership challenge that requires guiding teams through transformation while balancing both opportunities and risks.

AI as a Strategic Enabler, Not a Threat

Healthcare leaders must create an environment where AI is seen as a tool that enhances, rather than replaces, human capabilities. The American Medical Association (AMA) refers to AI in healthcare as "augmented intelligence," reinforcing its role in amplifying human expertise rather than replacing it. AI's impact is already being felt in areas such as:

- Administrative efficiency: generative AI can automate documentation, scheduling, and billing, reducing administrative burdens and allowing clinicians to focus on patient care.

- Clinical decision support: AI-powered diagnostic tools can assist physicians in identifying patterns and making data-driven decisions.

- Operational optimization: AI-driven automation can streamline workflows, optimize resource allocation, and improve hospital efficiency.

According to Forbes, AI's influence will "fundamentally change the day-to-day working lives of healthcare professionals,"[1] but human qualities like compassion and critical thinking will remain indispensable. The growing physician shortage shows no sign of letting up and based on research conducted by McKinsey.[2] Work-life balance and administrative and other tasks cutting into the time with patients are two of the leading causes for physicians leaving. Administrative tasks could be handled by AI, giving physicians more time with their patients and family. Again, the challenge for leaders is not whether AI should be adopted, but how to implement it responsibly, ethically, and effectively.

Building a Culture of AI Innovation

The success of AI in healthcare depends far more on the culture that surrounds it than the technology itself. Leaders must actively foster a culture of innovation where AI is not seen as a disruption but as an enabler of better care and improved efficiency.

Key Steps to Drive AI Innovation

    1. Prioritize education and training.

Equip teams with the knowledge and skills to use AI effectively through hands-on learning, mentorship, and upskilling programs.

    2. Encourage experimentation.

Establish AI pilot programs that allow employees to test, refine, and integrate AI solutions before full-scale deployment.

    3. Promote transparency and communication.

Address fears and misconceptions about AI by openly discussing its benefits, limitations, and ethical considerations.

    4. Invest in workforce development.

Provide ongoing training, ensuring that AI adoption enhances job roles rather than creating uncertainty.

    5. Develop AI governance frameworks.

Implement policies for ethical AI use, data privacy, and compliance to safeguard patient safety and regulatory integrity.

A recent Society for Human Resource Management article emphasizes that organizations that invest in upskilling their workforce will be best positioned for AI success.[3] AI-driven transformation requires more than technology adoption—it requires a workforce that is trained, empowered, and ready to embrace new capabilities.

The High Cost of Inaction: A Leadership Imperative

Failing to embrace AI is not a neutral decision—it is a risk. Healthcare organizations that lag in AI adoption risk becoming obsolete, much like past industry giants such as Blockbuster and Kodak, which failed to adapt to digital transformation. However, in healthcare, the stakes are far higher—lagging behind in AI adoption can directly impact patient outcomes, efficiency, and the financial sustainability of an organization. Moreover, healthcare AI comes with complex ethical, legal, and regulatory challenges. A Science Direct report warns that AI integration in healthcare requires careful oversight to address concerns such as:

- Patient safety and trust: it is critical to ensure AI-generated recommendations are accurate and evidence-based.

- Privacy and security: it is critical to protect patient data from breaches and misuse.

- Regulatory compliance: it is critical to adhere to frameworks like HIPAA, GDPR, and emerging AI governance standards.

Healthcare leaders must be proactive in shaping AI policies, not reactive. Failing to establish ethical and operational safeguards will create regulatory vulnerabilities and erode trust among patients and clinicians.

A Call to Action for Healthcare Leaders

AI is no longer an emerging trend—it is an essential tool for shaping the future of healthcare. The question is no longer "Should we adopt AI?" but "How do we implement AI effectively and responsibly?" To successfully lead AI transformation, healthcare leaders must:

- *Develop a clear AI strategy* that aligns with organizational goals.

- *Foster a culture of AI literacy* by investing in continuous learning and workforce development.

- *Ensure AI is an enabler, not a disruptor,* by integrating it into workflows where it adds tangible value.

- *Address ethical considerations head-on* through governance frameworks that prioritize patient safety and compliance.

- *Champion change management* by equipping staff with the tools and confidence to navigate AI adoption.

By viewing AI as a tool for empowerment rather than a threat, healthcare leaders position their organizations for long-term success. The biggest risk is not AI itself—it's failing to prepare for its impact. Those who embrace AI with strategy, governance, and a focus on people will lead the next era of healthcare innovation.

Envisioning the Future of Healthcare—Emergence of Artificial General Intelligence

It is the year 2035. A patient walks into an advanced emergency department with early signs of an undiagnosed condition. Within moments, an artificial general intelligence (AGI)-powered system synthesizes data from wearable sensors, genomic sequencing, and clinical records to generate a precise diagnosis and a personalized treatment plan. Ambient AI continuously monitors the patient's vital signs, while agentic AI optimizes staffing and resource allocation in real time, ensuring that care is delivered without delay.

This scenario is not a distant dream—it is a rapidly approaching reality. With breakthroughs increasing at an unprecedented pace and as AI is rapidly evolving from narrow AI to artificial general intelligence, healthcare is on the verge of a change in thinking. These advancements become very powerful when coupled with the AI-first transformation promise to improve care delivery, streamline administrative processes, and enable real-time decision-making on an unprecedented scale. Yet, as we stand on the threshold of this transformation, challenges related to governance, ethics, workforce adaptation, and cybersecurity remain critical.

Shaping the Future of Healthcare

AI offers healthcare leaders an unprecedented opportunity to transform their organizations. By moving beyond incremental improvements and embracing an AI-first mindset, healthcare systems can unlock exponential growth, improved patient outcomes, and operational excellence.

AI technologies are redefining healthcare by enhancing patient care, streamlining operations, and enabling data-driven insights. However, their full potential will only be realized through responsible adoption. Healthcare leaders must invest in governance, training, and collaboration to ensure that AI-driven innovations are ethical, equitable, and sustainable. By fostering collaboration between AI and healthcare professionals, investing in data governance, and educating patients about AI's role, healthcare organizations can unlock the full potential of this transformative era.

The future of healthcare is bright, but it requires thoughtful leadership to guide AI adoption in a way that prioritizes patient safety, data privacy, and clinical excellence.
—Dr. Michael Davis, Healthcare Futurist

The AI revolution as an inflection point, as a moment of change, an opportunity to offer new, more compassionate, and novel perspectives on care rather than to replicate existing problems, assumptions, and failures. Placing "care" back into healthcare would be a great starting point. This is not merely a technological shift—it is a call to reimagine how healthcare is delivered and experienced. As OpenAI CEO Sam Altman remarked, "This is the most interesting year in human history, except for all future years." The next decade promises to be the most transformative period in healthcare history, driven by the power of AI and human ingenuity.

Chapter Endnotes

1. https://www.mckinsey.com/industries/healthcare/our-insights/the-physician-shortage-isnt-going-anywhere

2. https://www.mckinsey.com/industries/healthcare/our-insights/the-physician-shortage-isnt-going-anywhere

3. https://www.shrm.org/topics-tools/news/hr-quarterly/7-trends-that-will-shape-hr-in-2025

About the Author

Victor Collins is Associate Director Data and AI at Impact Advisors, LLC, where he leads transformative initiatives at the intersection of artificial intelligence, automation, and healthcare strategy. With a proven track record in clinical analytics, intelligent automation, and enterprise data transformation, Victor partners with leading health systems to reimagine care delivery and operational performance through scalable, insight-driven technologies.

Victor holds a Professional Graduate Certificate in Data Science from Harvard Extension School, a certification in AI in Health Care from the MIT Sloan School of Management Executive Education program and East Carolina University School of Medicine. He brings a strong academic foundation to his leadership, blending technical proficiency with clinical insight.

A trusted advisor to healthcare executives and IT leaders, Victor specializes in designing human-centered AI solutions that are ethically grounded, operationally aligned, and regulatory compliant. His work spans predictive modeling, Epic EMR automation, and AI governance frameworks. Passionate about advancing equity, safety, and efficiency in healthcare, Victor is a recognized leader in applying emerging technologies to solve real-world clinical and administrative challenges.

Email: victor.collins@impact-advisors.com

LinkedIn: https://www.linkedin.com/in/victor-collins-7374138/

About the Author

CHAPTER 8

AI: THE SILK ROAD AND THE NEXT GREAT TRADE REVOLUTION

By Lisa Goodhand
Founder, China Blueprint Consultants; AI Specialist
Sydney, Australia

The new Silk Road is digital, and AI is its backbone.

When I first arrived in China in the early '90s, I had no idea I was stepping into a world on the brink of extraordinary change. I came to study Chinese, immersing myself in a culture vastly different from my own, captivated by the energy, the pace, and the sheer scale of possibility. But what I didn't realise was that I was witnessing the final moments of an era before technology would redefine global trade.

Back then, international business moved at the speed of trust. Deals were sealed over endless negotiations and copious amounts of green tea. Fax machines carried contracts across borders, and handwritten orders tracked shipments that took months to arrive.

Artificial intelligence (AI)? That was science fiction, not business. Success depended on who you knew—*guanxi*, as the Chinese call it—how well you could navigate bureaucracy and your ability to build relationships across the ocean.

Fast-forward to today, and trade has transformed before our eyes. The bustling trade fairs and agent negotiations of the past have been replaced by AI-powered platforms that automate transactions, predict market demand, and remove language barriers in real time. It wasn't just the mechanics of trade that changed; the very foundations of global commerce were being rewritten.

This transformation shaped my own career in international trade, as I found my passion in helping small businesses navigate this changing environment. Having lived through China's meteoric rise, I've had a front-row seat to the incredible ways AI has rewritten the rules—compressing timelines, eliminating inefficiencies, and creating opportunities that once seemed impossible for smaller players.

Much like the ancient Silk Road connected merchants across continents, AI has become the invisible infrastructure of modern trade. But where merchants once braved unpredictable terrain and weathered months-long journeys, today, businesses operate on digital highways. Algorithms negotiate deals. AI anticipates market shifts. And supply chains move at speeds that would have been unimaginable when I first set foot in China. The fundamentals of trade—trust, adaptability, and timing—are still there, but the tools have evolved beyond recognition.

What has impressed me the most is that the balance of power in trade has shifted. No longer is market access controlled by empires or corporate giants. Today, anyone with an internet connection can participate. AI and automation have been great equalisers. They've levelled the playing field, allowing small businesses and solo entrepreneurs to compete without warehouses, inventory, or the need to mortgage their houses in exchange for bundles of cash. What remains is opportunity.

Trade is no longer about ownership—it's about intelligence. The new Silk Road is a digital one, and in this era, adaptability isn't just an advantage: it's the price of entry.

China's AI-Driven Trade Transformation

Watching the rise of AI and digital trade from Australia over the past three decades has revealed a profound shift in global commerce—one where AI isn't just shifting power, but redefining access. This transformation is creating new opportunities for small and medium businesses, not just in Australia, but worldwide.

With AI-driven analytics, cross-border e-commerce, and seamless digital transactions, barriers that once favoured only large corporations are disappearing. Australian businesses—making up around 98% of the economy—are now tapping into global markets with unprecedented ease, reaching consumers in ways that were once unimaginable. From streamlining supply chains to predicting market trends, AI is levelling the playing field, making international trade more accessible for businesses of all sizes.

My Marco Polo Adventure

Like Marco Polo in his youth, I arrived in China, wide-eyed, stepping into a world I had only imagined. At that time, I was studying in Hangzhou—the same city a young Chinese English teacher named Jack Ma called home. He was earning just 100 to 120 yuan a month (about $20 USD) and had faced a string of rejections, including one from KFC. Like most people in China at that time, he had likely never heard of the internet. But when he did, he didn't just embrace it. He built Alibaba, the company behind Tmall (B2C) and Taobao (C2C), and helped shape one of the most significant digital revolutions in history.

Over the next decade, China's cash-based, street-market economy leapfrogged credit cards entirely, diving headfirst into the mobile payments era. E-commerce wasn't just a convenience—it became the backbone of business, both domestic and international. Platforms under Alibaba, along with others like JD.com and Dangdang, weren't merely facilitating transactions; they were reinventing commerce, integrating AI-driven recommendations, automated logistics, and livestream shopping into a seamless digital ecosystem. By the early

2010s, China wasn't just participating in digital commerce—it was defining it, scaling at a pace unmatched anywhere else.

While many countries took an incremental approach to e-commerce, China built a system from the ground up. Alibaba.com dominated B2B trade, linking businesses worldwide to Chinese manufacturers, while AliExpress expanded into B2C, allowing international consumers to purchase directly from Chinese sellers. By 2014, Tmall Global took this further, enabling foreign brands to reach Chinese consumers without needing a local presence. This shift positioned China as the driving force in AI-powered cross-border commerce, setting new standards in logistics, digital payments, and customer engagement.

Meanwhile, Australia found itself at a crossroads. Unlike China, it hadn't vaulted into AI-powered commerce, and digital retail adoption remained relatively slow. Traditional brick-and-mortar businesses continued to dominate, with many small enterprises viewing online sales as an optional extra rather than the inevitable future of trade. Yet, as global commerce became increasingly digitised, Australian businesses faced a choice: adapt or risk being left behind.

Australia's Digital Awakening: COVID-19

Then came COVID-19. Practically overnight, Australia-China trade—one of the economy's key pillars—was thrown into chaos. For decades, businesses had relied on in-person deals, trade fairs, and overseas trips, to build trust and secure supply chains. Suddenly, borders closed, flights stopped, and business as usual vanished.

At first, companies scrambled to adapt, shifting online out of necessity, not strategy. Retailers rushed to launch e-commerce stores, brands pivoted to digital marketing, and food delivery apps boomed as restaurants fought to survive. But few realised that AI was already working behind the scenes, driving this transformation.

As Australia found its footing post-pandemic, AI became the real catalyst for change. Businesses that had rushed into digital tools started turning to AI-powered solutions to optimise operations, increase efficiency, and expand into new markets. AI wasn't just helping them survive. It was pushing them forward.

Take Australian wine exporters, for example. Before COVID-19, selling to China required navigating trade fairs, personal introductions, and lengthy negotiations. Post-pandemic, AI-powered market analysis tools helped wineries track real-time consumer demand, automate marketing to target Chinese buyers, and streamline payments with AI-driven financial platforms.

Just as Australia was adapting to this new digital reality, AI accelerated the transformation, pushing trade into a data-driven, automated era.

And in December 2023, Jack Ma cemented this reality with his statement, "The era of AI-powered e-commerce has just begun." His words weren't just a prediction. They were a wake-up call. AI isn't just reshaping trade. It's creating a new digital Silk Road, bridging markets, translating languages, predicting consumer demand, and automating logistics in ways that would have seemed like science fiction just a few years ago.

It's fun to imagine what Marco Polo would think if he saw it now. I'd bet he'd trade his caravan messengers for a smartphone and a WiFi connection in a heartbeat.

AI and the New Era of Trade: How Small Businesses Are Now Going Global

E-commerce has revolutionised global trade, but for small businesses, breaking into international markets like China has always been an uphill battle. Complex regulations, cost, cultural barriers, and digital ecosystems vastly different from the West have made expansion challenging—until now. AI is changing everything.

In my time, I've watched AI tear down five major barriers in cross-border e-commerce, turning what used to feel impossible into a real, tangible opportunity. It's completely changed the game: helping businesses understand new markets (research), build trust and real connections (branding), create online stores that can sell globally (website development), reach and engage consumers—Chinese ones in my case (social media marketing)—and keep customers coming back with seamless, personalised support, all without ever leaving home (CRM).

AI isn't just making global e-commerce easier. It's making it smarter. From decoding consumer behaviour and refining brand positioning, to navigating China's unique digital landscape and personalising customer experiences at scale, AI is opening doors that were once firmly shut.

So how exactly is AI making this possible? Let's break it down.

Breaking Borders with Research

AI is reshaping global trade by eliminating barriers that once made cross-border business slow, expensive, and inaccessible. AI-powered research tools can now scan Chinese websites, social media, and e-commerce platforms in real time, turning months of manual work into instant insights on market trends and consumer sentiment.

For businesses targeting China's billion-strong market, this is a game-changer. AI enables smaller brands to act fast, make data-driven decisions, and adapt instantly—something once only possible with big budgets and teams on the ground.

China's retail giants, like Shein and Temu, have mastered AI-driven predictive analytics, using data to anticipate trends and optimise inventory. Now, demand forecasting software is making this technology accessible to smaller businesses, helping them stay ahead of market shifts instead of reacting to them.

AI-Powered Branding: Breaking Language Barriers and Building Global Trust

In culturally complex markets like China, branding can make or break success. A slogan or campaign that works in Australia might fall flat—or worse, offend—Chinese consumers. In the past, a single misstep could cost years of progress.

AI is eliminating that risk. Sentiment analysis software tests messaging in real time, ensuring it resonates with local audiences. Neural machine translation (NMT) tools refine tone and intent, adapting branding to feel authentic, not foreign. Natural language processing (NLP) goes even further, helping brands fine-tune product names, slogans, and campaigns for cultural nuance—without costly trial and error.

But branding in China is about more than messaging. It's about trust. Once built in bustling markets, today, trust is forged online through influencers and livestreaming. AI is now opening this once-exclusive space to foreign brands. AI-powered virtual influencers—fluent in Mandarin, available 24/7, and endlessly adaptable—engage audiences in real time, selling products without the need for a local presence.

Luxury brands in China are leading this shift, reducing their physical store footprint and embracing AI-powered omnichannel retail. Companies like Burberry and Gucci are leveraging AI-driven personalisation, virtual shopping assistants, and smart analytics to enhance the customer experience across digital and physical touchpoints. Live commerce and AI-backed digital experiences are now central to their China strategy, reinforcing how AI is becoming indispensable in engaging the next generation of Chinese luxury consumers.

The result? A direct, scalable entry into China's massive consumer market—no physical presence required. AI isn't just helping brands fit in; it's giving them an edge like never before.

Websites Unlocking Global Sales

For years, selling internationally meant playing by third-party marketplace rules and losing control over customer data, visibility, and brand identity. But that's changing fast. AI-powered websites are putting businesses back in the driver's seat, making it easier than ever to build and run a global e-commerce operation without relying on external platforms.

These websites aren't just smarter. They're borderless. AI now translates content, adjusts layouts, and optimises user experiences in real time. AI-driven drop shipping and fulfilment solutions also remove the need for warehouses, enabling businesses to scale globally without significant overhead.

Perfect Diary, a Chinese cosmetics brand, used AI-powered insights and social commerce to scale beyond China, expanding into Southeast Asia and other global markets. Conversely, the New Zealand-based Manuka honey brand, Comvita, leveraged AI-driven e-commerce tools and WeChat integration to enter China, where AI-powered personalisation helped it localise its marketing, automate customer interactions, and drive direct engagement with Chinese consumers.

When it comes to payments, navigating global transactions has always been a challenge—from fluctuating exchange rates to regional payment preferences and security risks. AI is changing that. AI-powered payment gateways, real-time currency conversion, and compliance tools are making cross-border transactions smoother, safer, and more accessible than ever.

Take China, for example, with 884 million online shoppers, yet credit cards are rarely used. Instead, the digital payment platforms WeChat Pay and Alipay dominate, backed by AI-driven fraud detection and real-time risk assessment. The good news? Businesses no longer need complicated workarounds to sell in China. AI now makes it possible to integrate seamlessly, ensuring fast, secure transactions in a market that once felt out of reach.

But AI isn't stopping there. It's optimising pricing, predicting trends, and personalising shopping experiences at scale. The same AI-driven strategies that power Shein and Temu are now available to smaller brands, giving them a real shot at competing globally—without a massive budget.

AI, Social Commerce, and the New Global Marketing Playbook

Social media is no longer just about connection. It's the new gateway to global e-commerce. AI-driven platforms are removing barriers that once kept international brands locked out, offering real-time translation, automated trend analysis, personalised content, and precision targeting. Businesses no longer need massive budgets or teams on the ground to break into new markets. AI is making global brand expansion faster, smarter, and more accessible than ever.

In China, where *guanxi* (personal relationships) once dictated business success, social commerce has taken over. Platforms like WeChat, Xiaohongshu (RedNote), and Douyin (China's TikTok) are driving e-commerce through influencer marketing, live shopping, and AI-driven recommendations. Until recently, language barriers, cultural nuances, and unfamiliar algorithms made these platforms nearly impossible for foreign businesses to navigate. Now, AI is bridging the gap, helping brands localise their messaging, connect with engaged audiences, and scale their presence seamlessly.

Even global celebrities are taking notice. In 2023, Kim Kardashian joined a Xiaohongshu livestream to promote SKIMS, tapping into China's massive consumer base. But this isn't just for celebrities. AI-powered algorithms are levelling the playing field, allowing smaller brands, influencers, and startups to compete internationally without the need for a massive following. Unlike traditional social media, where reach was dictated by follower count, AI prioritises relevance and engagement, ensuring the right content reaches the right audience.

And it's not just social media that's evolving. SEO has changed too. While traditional SEO is still important, it's often complex, expensive, and slow to deliver results. For small businesses, AI-powered social media marketing offers a faster, more accessible path to international sales. Search rankings are no longer just about keywords and backlinks. They're now shaped by social engagement, AI-driven content recommendations, and real-time user behaviour. AI fuels both search and social commerce, meaning businesses that integrate AI-powered social strategies into their global marketing will be the ones leading the next wave of cross-border e-commerce.

AI Is Redefining Global Trade—And Small Businesses Are Finally in the Game

For years, expanding into international markets felt impossible for small businesses. Even large corporations struggled with the cost of maintaining teams abroad. Just look at Airbnb, which shut down its domestic business in China in 2022, citing high operational costs and complex local requirements. If a company of that scale found it unsustainable, what chance did smaller businesses have? Without local teams, multilingual customer service, and complex logistics networks, competing globally was a challenge only big brands could afford. But now, businesses of all sizes are finding new ways to scale internationally without the overhead.

Amazon, on the other hand, has operated successfully in global markets by leveraging AI to automate customer service, logistics, and personalised shopping experiences. Its AI-powered customer support systems, chatbots, and predictive analytics handle millions of inquiries across languages and time zones, providing seamless, localised service without the need for human intervention. Today, those same capabilities—automated CRM, multilingual chatbots, and real-time customer sentiment analysis—are accessible to small businesses through scalable, cost-effective tools.

But AI's role in global trade goes far beyond customer support. It's also transforming logistics and supply chain management, removing bottlenecks that once made international expansion difficult. AI-powered inventory systems and smart tracking solutions now handle demand forecasting, automated stock replenishment, and real-time delivery updates, helping businesses manage operations more efficiently, while improving the post-purchase experience for customers worldwide.

This is just the tip of the iceberg. AI isn't just making global trade possible. It's making it effortless. Want an international expansion plan for your next presentation? Need ideas on marketing in China? Curious about the biggest buyers of skincare in Vietnam? Try typing it into ChatGPT—but be careful with that next sip of coffee… your jaw might hit the floor.

Welcome to the future—where AI just became your best employee.

AI and the New Era of Learning

AI isn't just streamlining trade. It's transforming how businesses learn, adapt, and grow. Like Marco Polo, small businesses should approach new markets with a learning mindset, embracing discovery rather than fearing the unknown. His success wasn't just in the goods he brought back, but in the knowledge he acquired, shaping trade for generations.

AI is a tool for exploration, but it's not infallible. It lacks human intuition, makes mistakes, and can misinterpret cultural nuance. The real advantage lies in combining AI with human expertise—leveraging technology while staying agile, informed, and culturally aware.

This isn't just another shift in trade—it's an opportunity. AI is breaking barriers of language, geography, and scale, creating new pathways for businesses to expand globally. The companies that master its power—from AI-driven research and branding to digital storefronts, social commerce, and customer engagement—won't just survive this revolution. They'll lead it.

About the Author

Lisa Goodhand is a dynamic entrepreneur, digital strategist, and specialist in China-Australia trade relations. Since founding China Blueprint Consultants in 2008, she has helped Australian businesses successfully enter and navigate the complex Chinese market. She speaks Chinese and has a deep understanding of Chinese social media platforms such as Xiaohongshu (RedNote), WeChat, and Douyin, specialising in cost-effective digital strategies that drive brand awareness and sales.

Her expertise spans market research, branding, ecommerce, and influencer collaborations, with a focus on empowering SMEs to expand internationally. Passionate about bridging cultural and business gaps, Lisa frequently shares insights on cross-border commerce, emerging trends, and AI-enhanced marketing. She is a sought-after speaker and trusted advisor for businesses looking to grow in China.

Under her leadership, China Blueprint Consultants continues to pioneer innovative, results-driven strategies that make international expansion accessible, scalable, and profitable for Australian brands.

Email: lisa@chinablueprint.com.au

Website: https://www.chinablueprint.com.au/

CHAPTER 9

UPGRADED ALGORITHMS— DOWNGRADED HUMANS

By Basil Hartzoulakis, PhD
Founder & Director, Ethical Product Management
Cambridge, England, United Kingdom

> *Whenever I run into a problem I cannot solve, I always make it bigger. I can never solve it by trying to make it smaller, but if I make it big enough, I can begin to see the outlines of a solution.*
>
> —Dwight D. Eisenhower,
> World War II, Allied Supreme Commander

The proliferation of harmful AI applications is likely to persist unless governments implement internationally binding regulations that designate such uses of AI as "unethical." Only democratic societies, rooted in human values such as love, equality, family, empathy, compassion, collaboration, rule of law, and community, can effectively harness AI to foster prosperity while mitigating downside risks of inequality, totalitarianism, conflict, environmental degradation, and poverty.

The Fear of AI

What is AI? Is it an "alien" intelligence? "Who" beat Gary Kasparov in 1997? Are Terminator-like machines about to take over your job or destroy humanity? Can we stop them?

Gary Kasparov, world chess champion, lost his second game against IBM's Deep Blue supercomputer in a tightly contested and well-publicized game in 1997. Deep Blue was a state-of-the-art system; chips designed for running optimized chess algorithms, databases with past games, and real-time input from other chess grandmasters. Add to the mix, the combined experience and skills of 20 engineers and mathematicians. So, Kasparov was not facing "someone." He was playing a "tug of war" against a group of humans who used technology to align their intelligence against him. Deep Blue was just the "rope". Have things changed since then?

Today's news discusses two categories of artificial intelligence models: generative AI and large language models (LLMs). Both types share common algorithmic elements and are suitable for different applications. The former deals better with pattern recognition (e.g., visual or auditory) where the latter deals with text. They are more sophisticated than Deep Blue and much more expensive to train. The models use human input, break it down to pieces, and then combine the elements to perform a task. As an algorithm they are just a set of instructions on how to contextualize and simulate human activities.

Now that we understand how they work, we can ignore qualifications as "alien" or "dangerous." AI will not achieve consciousness or ever be creative. Can the models recognize patterns in certain types of data? Yes, they can. Can the models identify new inferences, be creative, or acquire a purpose? No, they cannot.

Still, their ability to write "good" prose, summarize documents, and write a risk assessment policy draft in seconds is impressive and causes excitement, disquiet, and alarm in equal measures. To assess the situation accurately and gauge our options, we need to take a longer view of technology.

Humans have strived to make their lives easier since the dawn of civilization. Technology and tools have been the means to this end. We need a predictable food supply, we invent agriculture. We want to honor gods or build a tomb that will last "forever", we develop mathematics and architecture. We want to occupy the land of our neighbors, we build better weapons. We want wealth and variety, we invent money, sailing, and trade.

Still, until the 13th century the pace of technological innovation was slow. Applied disciplines like medicine (health) and mathematics (architecture) advanced at a faster pace but soon reached limitations (e.g., you cannot operate without antibiotics). Three observations from antiquity are relevant to our AI challenge.

Technology in the Ancient World

Technical innovation was a secondary pursuit, sometimes even frowned upon. Human success was a balancing act between social, religious, philosophical, and political norms. A skilled mathematician, tradesman, farmer, or sailor would conceive an idea and build it, that was it. Technological innovation was a fun side activity. A notable example is gunpowder, invented during the effort to discover the elixir of life.

Powerful rulers used technology to grow their empires and wealth—no surprises here. Famous examples are silk production, kept as a trade secret for centuries, and the Byzantine "liquid fire" weapon whose recipe was so well protected that it got lost! The pursuit of scientific knowledge and innovation, for the benefit of the wider society, did occur under the leadership of benevolent rulers, but it was fragile and short lived. Ancient Athens, Alexandria in Egypt under the Ptolemies, the Abbasid Caliphate, certain Medieval monasteries, and early universities are the main examples.

The "Growth Flywheel"

From the 14th century onwards, humanity went into overdrive. How did we do it? First, we changed our baseline of belief and value systems. Aristotelian thought started competing with religious doctrine. Universities sprang everywhere and the newly educated middle classes considered Christianity values too restrictive. Descartes declared "*Cogito ergo sum*—I think, therefore I am" (1637). A line of philosophers brought us to Nietzsche who declared, "God is dead", and provided the foundation for relativism and perspectivism. We rejected the notion that there are objective, universal truths and moral principles that apply to all people, regardless of context or perspective. Values from philosophy, arts, religion, or community tradition were deprioritized. Logical thought and intelligence became the only game in town. Technologists and philosophers now had permission to reduce human existence and experience to a series of structured logical thoughts and mathematical equations—the seed of AI. Parliamentary democracy, in a national state context, became the political system expressing our new understanding of the world and our reliance on logic and science. Three innovations completed the flywheel: military technologies (based on gunpowder), the printing press, and stakeholder capitalism. The East India Company was founded on December 31, 1600.

The modern era kicked off with a blast, printing, and trading. The impact of our inventions soon became global, so did the magnitude of the positive and negative, always unpredictable, side effects. Nations or corporations, yielding technological advantages, accumulated power and resources and shared them with the emerging middle classes controlling the financial system. Wealth accumulation and colonialism became synonymous with progress. Corporations keen to make profit for investors followed through and provided financial support.

In the 17th century (*Principia Mathematica*) we adopted the doctrine of science, logic, and technological innovation as the best way to ensure that society is healthy, vibrant, progressive, and democratic. The doctrine underpins the AI revolution's narrative today and assumes unlimited progress and infinite resources.

The "growth flywheel" does work. We witnessed an exponential increase in food production and the human population. Sanitation, antibiotics, vaccines, and medical advances reduced child mortality and doubled our life span. Western countries used their wealth to build a welfare state that secured jobs and protected citizens and families from misfortune or illness. After World War Two, progressive tax policies were adopted, and workers were granted protection by the law to negotiate fair wages and improved working conditions with their employers.

Were these benefits equitably distributed across the globe? Wealth increases in one part of the globe resulted in poverty, famine, and worse living conditions elsewhere. Communities with no access to technology were judged unworthy, enslaved, forced to submission, or even eliminated.

The first Industrial Revolution (18th to 19th century) pushed into poverty and famine whole regions (India, China, Africa, Southeast Asia). Living conditions for working people in the West were dire for 200 years. They improved drastically after World War Two with the New Deal and the welfare state. In parts of the world, especially in Africa, this collapse in living standards persists.

The second Industrial Revolution (20th century) increased the global industrial capacity by a hundred-fold. We got cars, fridges and washing machines, gadgets, consumerism, and fast fashion. Our binge in oil was extraordinary. In 50 years (since 1970) we turned our atmosphere and oceans into dumping grounds for carbon dioxide, plastic pollution, and chemicals. We destabilized the climate and put at risk the ecosystem balance that allowed humanity to evolve in the first place. Cleaning up the mess burdens the public purse but not the companies or investors who developed the products in the first place (e.g., plastic pollution, electronic devices).

Internet and Global Connectivity

The NET is a waste of time, and that's exactly what's right about it.
—William Gibson, 1996

The internet revolutionized communication but also exposed societies to unprecedented risks with no other measurable societal benefits. Inadequate data protection has allowed states and malicious organizations to create new forms of crime and terrorism. The early promise of new public spaces morphed into echo chambers where misinformation and lies proliferate with astonishing speed. Dictatorial governments found a new way to control the news feed, the education, and the sentiment of their citizens. The internet became the best "Big Brother" enabler.

Social Media

These platforms enable a form of communication between users devoid of any humane character. Scrolling causes brain development problems including attention deficit and learning impairments. Research on the negative impacts on children's health is so devastating that it should have moved our governments to action long ago. We were quick to lock the population down during the COVID-19 pandemic, but we seem unable to act to prevent well-documented harm to our children. Social media uses AI to deliver profits and encourage addictive behavior for children and adults.

Despite positive average numbers in the last 50 years, we have witnessed a drop in the living standards for ordinary citizens. Inequality has been increasing, innovation has slowed down, and war and conflicts have doubled affecting one in eight people around the globe. Interestingly, governments and citizens are getting poorer and indebted. Recent calculations showed that most Western governments do not own enough wealth to pay off their debts. As a result, they have no resources to serve and protect the vulnerable. Weak governments are unable to collaborate internationally and control giant global corporations. The "growth flywheel" has helped us live longer, feed ourselves, and get our children educated, but the cost in terms of inequality and environmental degradation is difficult to defend.

Current State of AI and Key Trends

The future is here—it's just not evenly distributed.
—William Gibson, 2003

What impacts to expect imminently by the AI revolution? AI applications fall into four categories with different impact profiles on our lives.

1. Modelling

AI algorithms will help scientific R&D, materials sciences, and process optimization in most industries. New products will take years to market, with a focus on medicine, energy, and climate tech.

2. Assistive

Applications are available for personal use (e.g., ChatGPT) or companywide implementations. Industries affected include call center operators, drivers, and warehouses. Knowledge workers and professionals will also experience changes. Job losses, gig-economy work patterns and loss of income are all expected outcomes. Corporations will get powerful tools to promote automation and make humans work more like machines. A new paradigm of work will enable algorithms to decide and implement efficiency improvements in real time. Financial gains will be siphoned to the AI tool providers and shareholders. No benefits will trickle down to workers or other professionals.

The key challenges will be to protect your privacy rights at work, stay engaged in your workplace with opportunities to learn and grow, and compete in a healthy job market with no monopolies. One area of positive promise is the deployment of assistive AI as a decentralized agentic service owned and trained by individuals and small and medium enterprises (SMEs). Affordable, bespoke AI agents will be key for adoption and can prevent "winner takes it all" scenarios.

3. Behavioral Modification

Behavioral modification applications are everywhere. Digital platforms constantly profile us and adjust the service to increase engagement or revenue. This application serves equally digital platforms and oppressive governments. The use of automated decision systems falls into this category. Healthcare applications and banking are two critical markets. It will be interesting to see how AI systems will cope with the strict data privacy and regulatory requirements in these sectors.

4. Military and Autonomous Systems

The final category describes military applications and autonomous systems deployed in emergencies but also in a conflict. This sector has recently attracted three of the highest startups in recent valuations in the Silicon Valley, highlighting the increasing venture capital interest for AI application in warfare.

The "growth flywheel" doctrine is still strong. Governments and corporations mostly agree on a low regulatory environment with profits and technological supremacy as the key drivers. Under the current political climate and level of debt, authorities cannot enforce measures that could help the workforce with the transition or protect vulnerable groups.

The Ideal Framework for AI Development

The internet and social media delivered a poisonous mix of consumerism and harm to our children and public institutions. Both technologies have created unfamiliar problems that need resolution. In this context, AI is a similar "high-risk, high-reward" bet. The risk is not that AI will become a sentient threat but that it will increase inequality and weaken further democratic institutions and global cooperation. What are the features of a resilient democracy that can control AI for the benefit of society and still allow companies to make a profit?

1. Citizenship

Citizens should have freedom, time, and financial support by the state to interact in community projects and participate in the commons. The relationship between citizens, groups, and authorities should be direct, not mediated by computer systems or commercial interests.

2. Corporate Responsibility

Corporations are making a profit and providing jobs, but should respect the context and the values of the society they are operating in. They should pay taxes and not interfere with the executive or the legislative aspects of governments.

3. Risk Management

Government has enshrined robust regulations in law to manage risks and mitigate the inevitable side effects of AI. Rules are also in place to hold corporations to account for damages arising from the use of their platforms and tools. International cooperation ensures a level global playing field for AI developers.

4. Wealth Distribution

A mechanism should be in place that allows an equitable sharing of the benefits and ensures resources are available to deal with undesirable social externalities.

Time for Action

The best way to predict the future is to create it!
—Abraham Lincoln

The AI revolution requires us to act with foresight and boldness. The aim is to protect our families, our communities, our income, and the environment.

1. Citizenship

First let us declare that AI, like every technology, has negative impacts that society needs to address. We need to educate ourselves on the concepts discussed here and adopt a framework of values that encourages community and collaboration above wealth and competition. There should be mechanisms in place (assemblies, direct democracy) where citizens can inform government policy on a regular basis, especially on fast-moving topics like AI or the climate emergency.

2. Regulatory Framework and Risk Management

Engage and support in public government initiatives that impose a framework for regulating AI. The EU approach to classify AI products in four groups is a good starting point ((1) unacceptable risk/prohibited, (2) high risk, (3) limited risk and (4) minimal and/or no risk). A long-standing example is the pharmaceutical industry and the evaluation steps a new drug goes through before patients can access it. Support government initiatives that create a level playing field for SMEs against larger corporations. The riskiest AI systems—defined by both capability and use cases—should face higher regulatory hurdles before they reach consumers. Children's education is a primary example, where evidence-based approaches should be in place.

Measures like placing liability for harms on developers, mandating reporting requirements, establishing legal protections, and providing incentives for responsible developers and deployers who meet risk management and reporting requirements can all promote safe AI development. Stakeholders, governments, corporations, and citizens must work together within a democratic framework to manage risks effectively. Addressing meta-problems such as inequality and environmental degradation is essential, rather than solely focusing on AI technologies.

3. Corporate Responsibility

Commercial organizations should have input but no influence over legislation. Laws should be in place to hold them to account if they ignore their obligations to society, trying to bypass regulation, avoid taxation, or lobby governments for favorable laws. As professionals we need to adopt an international ethical product framework for AI products. Software or platforms should adhere to a value system that is both lawful and compliant with ethical customs and beliefs of our users and stakeholders. Ethical product management should be the method for reducing externalities and increasing profitability at the design stage of every product.

4. Wealth Redistribution

Policies like progressive taxation or employee share schemes can help create a fairer economy. Governments should aim to manage economies with lower debt levels, as excessive debt limits future options and carries long-term consequences. A sustainable economic environment should enable people to prosper without relying on increasing personal or public debt.

AI is a great chance for humanity to revise the "growth flywheel" and infuse it again with core human values of love, empathy, community, and equality. We face a choice: align with finite values—money, status, power, and competition—or embrace infinite values—love, equality, family, community, compassion, collaboration, and environmental stewardship. The path we choose will shape the future of AI and its impact on society.

About the Author

Dr. Basil Hartzoulakis is a product management director, entrepreneur, and computational chemist with over 20 years of experience transforming deep tech and life science innovations into market-leading solutions. His expertise spans AI-driven SaaS, biotechnology, and scientific informatics, where he has consistently delivered high-impact, award-winning products.

Basil is passionate about ethical product development and applying deep technology to solve real-world problems. He is currently engaged in leadership and consulting opportunities in B2B SaaS, market strategy, and scaling product functions, where he helps teams build, launch, and scale transformative products.

Basil is a champion martial artist (National British, 2006) and Shaolin Kung Fu teacher (18 years). He has served as a volunteer coach and mentor. His wife and three children teach him daily lessons in love, humility, and acceptance. He meditates, goes to church, visits Crete, the land of his family, keeps chickens, and cooks traditional Cretan dishes for family and friends.

Email: basil@aiagents.co.uk

Website: www.ethicalproduct.co.uk

CHAPTER 10

THRIVING IN THE AGE OF HUMAN-CENTERED AI

By James L. Hutson, PhD
Multidisciplinary AI Researcher
Saint Charles, Missouri

> *Technology is best when it brings people together.*
> —Matt Mullenweg

Artificial intelligence seemed to burst into public consciousness overnight with the launch of OpenAI's ChatGPT in November 2022, a conversational tool so advanced it felt like talking to a knowledgeable and infinitely patient expert. This sudden spotlight, however, obscures the truth: AI has been quietly shaping our world for decades. From rudimentary chess programs of the 1950s to the algorithms that curate our social media feeds, recommend movies, and approve mortgage applications today, AI has been steadily weaving itself into the fabric of our daily lives. The difference now is that it is not just working behind the scenes; it is actively engaging with us, forcing society to reckon with its capabilities and consequences.

The rapid integration of AI into personal and professional spaces has sparked a whirlwind of possibilities. Industries as varied as healthcare, education, and creative arts are discovering how intelligent systems can amplify human capabilities. AI models now assist doctors in diagnosing illnesses with unprecedented accuracy, help teachers create personalized lesson plans, and empower artists to design works that blur the line between human and machine creativity. Yet, these advances come with undeniable risks: the potential for biases baked into AI algorithms, the threat of automation displacing jobs, and the ethical quandaries of allowing machines to make decisions that profoundly impact human lives.

At the heart of this moment lies an urgent question: how do we ensure that artificial intelligence serves humanity's best interests while preserving the values, ethics, and creativity that define us? To answer this, it is essential to recognize that there are many ways to approach AI. For the purposes of this chapter, however, we will focus on two distinct perspectives.

The first sees AI as an end unto itself—pursued for its intrinsic value, with goals such as achieving artificial general intelligence (AGI) or refining algorithms to ever-greater precision. This viewpoint often dominates the realm of computer science, where the primary aim is to push the boundaries of what AI can do, with little consideration of why it should be done or for whom. The focus here is on improving the tool for its own sake, driven by curiosity or the technical challenge.

The second perspective—and the one this chapter advocates—views AI not as a self-contained objective but as a means to serve humanity. From this vantage point, the question shifts away from how to make AI better for its own advancement to how AI can address real-world problems and enrich human lives. By reframing AI as a tool and placing people at the center of its purpose, we elevate its potential to solve pressing human challenges, foster creativity, and uphold ethical principles.

This shift defines the essence of human-centered AI (HCAI). Unlike efficiency-driven AI, which prioritizes optimizing tasks and outcomes regardless of context, HCAI insists that technological systems align with human needs and values. It prioritizes questions of relevance, equity, and impact over raw computational power or abstract performance benchmarks. HCAI does not seek to replace human ingenuity but to amplify and complement it.

As we explore the implications of HCAI, this chapter argues for a balanced approach to AI's role in society. It should neither be romanticized as a utopian savior nor vilified as a dystopian threat. Instead, AI must be embraced as a collaborative partner that, when guided by ethical, adaptive, and thoughtful strategies, empowers humanity to thrive in an era of profound transformation. By shifting the narrative from "How do we improve AI?" to "How can AI improve our lives?" we unlock its true potential—not as a machine that dictates the future but as a tool that enables a more just, creative, and humane world.

Embrace AI as a Collaborative Partner

In 2017, a survey revealed a striking sentiment among Americans: more people feared a robot taking their job than they feared death itself. This visceral reaction underscores the deep anxieties surrounding artificial intelligence, particularly its potential to disrupt industries and displace human workers. Media coverage and industry discourse often amplify these fears, highlighting scenarios where automation and AI systems render human roles obsolete. From assembly lines to customer service desks, stories abound of machines replacing tasks once thought to require uniquely human skills.

This unease, however, is not new. The fear of human-like machines performing human actions has existed since humanity's earliest stories. Legends like the ancient Greek myth of Talos, a giant bronze automaton, or Mary Shelley's *Frankenstein*, reflect a persistent wariness toward creations that could outstrip their creators. Yet while the form of these fears has evolved—from mechanical constructs to algorithms—the underlying anxiety remains the same: what happens when our tools surpass our control?

Rather than viewing AI as a competitor, reframing it as a collaborative partner offers a more constructive perspective. Unlike the anthropomorphic robots of science fiction, today's AI excels not in mimicking humans but in complementing our strengths. AI systems can analyze vast datasets at speeds unattainable for humans, assist in pattern recognition, and automate repetitive processes. These capabilities free individuals to focus on tasks that require creativity, emotional intelligence, and strategic thinking—qualities AI cannot replicate.

For example, in healthcare, AI helps doctors by analyzing medical images to identify potential health issues faster and with greater accuracy, enabling earlier interventions. In education, it personalizes learning experiences, identifying students' strengths and areas for growth, while teachers concentrate on fostering critical thinking and interpersonal skills. In business, AI streamlines operations and provides actionable insights, allowing leaders to focus on innovation and human-centered strategies.

To embrace AI as a partner, society must shift its narrative. AI is not a rival to humanity but a tool designed to augment human ingenuity. By understanding and leveraging its strengths, humans can direct AI to address challenges—from climate change to resource distribution—that require the unique combination of computational power and human creativity. The key lies in fostering collaboration, not competition, between humans and their technological creations.

Through reframing artificial intelligence as a collaborator rather than a threat, individuals and organizations can unlock its full potential, using it to innovate, solve problems, and improve lives. This mindset shift transforms AI from a symbol of fear into an agent of empowerment, fundamentally reshaping the way humanity engages with one of its most transformative inventions.

However, much of the discourse surrounding AI in the workplace remains steeped in negativity. News stories often emphasize a dystopian narrative, portraying workers who use AI as complicit in automating or offloading their roles, contributing to the erosion of job security. This framing overlooks the practical and creative ways in which AI is actually being adopted by professionals. Far from surrendering their roles to machines, employees are leveraging AI to tackle the repetitive and uncreative aspects of their jobs, enabling them to focus on work that requires uniquely human insight and creativity.

Emerging research underscores this trend, revealing that AI tools are empowering workers rather than replacing them. For instance, employees frequently use transcription tools like Otter.ai to automate the tedious task of taking meeting minutes. Instead of spending hours typing notes and creating action plans post-meeting, workers can rely on AI to generate accurate transcripts in real time, freeing their schedules for strategic thinking or creative problem-solving. Similarly, generative AI platforms like Claude are being used to spark idea generation. Professionals can brainstorm effectively, using these tools to produce foundational concepts that they can refine and develop further.

The implications are profound. Workers are not being displaced by AI; they are being equipped with tools that amplify their efficiency and enhance their job satisfaction. By automating mundane tasks, such as data entry or preliminary research, AI enables professionals to concentrate on areas where human skills shine: fostering relationships, driving innovation, and crafting nuanced solutions to complex challenges.

This approach demonstrates the symbiosis between humans and AI: technology enhances productivity by taking on tasks that were once time-consuming or monotonous, while humans retain their agency and focus on creative and strategic priorities. Rather than fearing a future where AI usurps human roles, this emerging paradigm suggests a workplace where humans and machines collaborate seamlessly to achieve outcomes neither could accomplish alone.

Such reframing is critical for shifting public perceptions of AI. By emphasizing its role in augmenting human capability, society can move beyond narratives of displacement and embrace AI as a tool for empowerment. This not only helps individuals to better understand and adopt these technologies but also encourages organizations to innovate responsibly, ensuring AI serves as a catalyst for growth, creativity, and shared success.

Ethical Responsibility and Navigating AI Regulations

As AI becomes increasingly integral to society, the need for ethical responsibility and effective regulation has never been more urgent. Ensuring that AI systems serve humanity equitably requires a commitment to principles such as fairness, accountability, and transparency. These frameworks act as guardrails to prevent misuse and mitigate harm, particularly in sensitive areas such as hiring, policing, and lending, where algorithmic bias can have far-reaching consequences.

Bias in AI systems often arises from the data they are trained on. For example, hiring algorithms trained on historical workforce data have been shown to replicate and even exacerbate discriminatory practices, such as undervaluing applicants from underrepresented groups. Similarly, predictive policing systems that rely on biased crime data have led to disproportionate surveillance of minority communities, compounding systemic inequities. In lending, algorithms can unintentionally penalize certain demographics due to historical patterns of credit access, reinforcing economic disparities. These examples highlight how AI, when deployed without ethical oversight, can perpetuate injustices rather than alleviate them.

Addressing these challenges demands a collective effort. Developers must prioritize inclusive data practices and implement rigorous bias detection mechanisms. Businesses should integrate ethical considerations into their AI strategies, fostering cultures of accountability and transparency. Policymakers, meanwhile, have a critical role in crafting legislation that balances innovation with protection, ensuring that AI systems are fair, safe, and aligned with societal values. Together, these stakeholders must share the responsibility of steering AI development toward equitable outcomes.

Global regulatory landscapes are beginning to reflect this imperative. In the European Union, the proposed AI Act seeks to categorize AI applications by risk level, imposing stricter regulations on high-risk uses such as biometric surveillance. The United States has introduced frameworks like the National AI Initiative Act, which aims to promote innovation while addressing ethical and safety concerns. Around the world, nations are grappling with the dual challenge of fostering AI's transformative potential and safeguarding against its risks.

For businesses, navigating these emerging regulations is not only a compliance issue but also a strategic priority. Staying informed about legislative developments is essential, as is proactively developing internal governance policies that align with regulatory expectations. Companies can also play an active role in shaping policy by engaging with regulators, sharing industry insights, and advocating for practical solutions that reflect real-world challenges.

Ultimately, ethical responsibility and regulatory navigation are two sides of the same coin. Both require a forward-thinking approach that anticipates potential pitfalls and works to address them before they arise. By embedding ethical principles into AI design and deployment and by actively engaging with regulatory frameworks, organizations can not only mitigate risks but also build trust with stakeholders and create systems that truly serve the common good. In doing so, they contribute to a future where AI drives progress without compromising fairness, accountability, or transparency.

Adopting an Adaptive Mindset for Continuous Learning

There is a flipside to ethical and responsible considerations in AI adoption: they have often fueled inertia, particularly in fields like education. Fear of the unknown and the rapid pace of technological evolution have led to paralysis, with some stakeholders leaning heavily on ethical concerns as a way to resist the technology altogether. While ethics is undeniably important, it should not prevent us from preparing for a future where tools like AI will be integral to the daily lives of Gen Alpha—the next generation who will rely on these systems to augment their abilities and navigate an increasingly complex world.

What truly worries me is the widening digital divide exacerbated by this reluctance. Communities already disadvantaged by a lack of access to technology face compounding inequities as AI becomes a baseline requirement for success. Imagine the disparity between a school district with robust WiFi and another struggling with connectivity issues. If we then add AI reliance to the mix, the gap becomes insurmountable. Addressing this issue requires a dual focus: ensuring equitable access to technology and fostering an adaptive mindset that enables all individuals to thrive in an AI-driven world.

An adaptive mindset is critical because the rapid evolution of AI demands lifelong learning and a willingness to embrace change. Skills once considered essential may soon be augmented—or even replaced—by AI, requiring a pivot to roles that emphasize human creativity, critical thinking, and problem-solving. Workers across industries need to develop agility, curiosity, and resilience to keep pace. Tools like Coursera, LinkedIn Learning, and AI-specific platforms such as Fast.ai offer accessible ways to build skills in areas like machine learning, data analysis, and AI ethics. For educators, professional development programs tailored to integrating AI in the classroom are increasingly available, helping them prepare students for an AI-enhanced future.

Early adoption strategies are equally important for organizations and individuals aiming to stay competitive. Practical tools already exist to streamline workflows and enhance productivity. For example, small businesses can use free or low-cost AI solutions like Otter.ai to transcribe meetings or Canva's AI-driven design features to create marketing materials. Educators can leverage platforms like Quizlet or Khan Academy's AI-driven learning paths to customize student engagement. These tools not only save time but also enable users to focus on more creative and strategic tasks.

To pilot AI effectively, start small and scale thoughtfully. Identify repetitive tasks that consume time and explore how automation can alleviate them. Develop clear objectives and metrics to measure AI's impact while maintaining a human-centered focus. Collaborate across teams to ensure that AI solutions align with organizational or educational goals. By minimizing risks and maximizing benefits, this approach fosters confidence and builds momentum for broader adoption.

Ethical responsibility and practical adoption are not opposing forces but complementary imperatives. By addressing ethical concerns without allowing them to paralyze action, communities can prepare for the realities of the future while narrowing the digital divide. This dual approach ensures that AI is not merely a privilege for the well-connected but a tool that empowers all individuals to succeed. The stakes are high, but the rewards of taking proactive, equitable, and adaptive steps toward AI integration are far greater.

The Imperative of Metacognition and Lifelong Learning

In an era where artificial intelligence evolves at breakneck speed, the greatest skill we can cultivate is metacognition: learning how we learn, so we can do it better and faster. This ability is essential not just for individuals but also for the next generation—Gen Alpha—who will navigate a world fundamentally transformed by AI. Predictions suggest they will experience up to ten major career changes in their lifetimes. To thrive in this dynamic environment, they must embrace lifelong learning as a foundational principle, and so must we.

Upskilling is no longer optional; it is an ongoing necessity. AI is not only transforming industries but also redefining the skills required to succeed within them. The key trait distinguishing those who lead innovation and progress is an open mindset, often referred to as a growth mindset. My research consistently shows that individuals who are out ahead—pioneering new ideas, processes, and technologies—share a willingness to adapt, learn, and iterate. They view challenges not as barriers but as opportunities to improve and grow.

Adopting this mindset begins with understanding that learning is a continuous process. Agility, curiosity, and resilience are the cornerstones of this approach. Agility allows individuals to pivot as industries change; curiosity drives them to explore and master new tools; and resilience ensures they persevere through inevitable setbacks. Developing these traits empowers both individuals and organizations to stay competitive in an AI-driven world.

Fortunately, resources for upskilling in AI are more accessible than ever. Platforms such as Fast.ai and Google's Machine Learning Crash Course offer free or low-cost opportunities to build technical skills, while broader platforms like Coursera and LinkedIn Learning provide comprehensive courses tailored to a variety of industries. For those in non-technical roles, tools like ChatGPT and Canva can help demystify AI and provide immediate, practical applications in everyday workflows. Educators, too, can access professional development programs to integrate AI into classrooms, preparing students for an augmented future.

Beyond acquiring technical skills, fostering metacognition—the ability to reflect on how we learn—amplifies the effectiveness of lifelong learning. Teaching children and adults alike to assess their learning strategies, recognize areas for improvement, and experiment with new approaches ensures that they remain adaptable. For example, encouraging students to reflect on what methods help them understand complex topics or prompting professionals to evaluate which tools most effectively enhance productivity embeds a culture of continuous improvement.

The rapid pace of AI development will demand this mindset more urgently than ever. Consider how industries are adopting AI to automate repetitive tasks, analyze data, and even create content. For individuals to thrive amidst these changes, they must not only learn how to use these tools but also reimagine how their roles can evolve alongside technology. Early adopters who approach AI with an open and adaptive mindset are already demonstrating how these technologies can drive innovation and unlock new opportunities.

In equipping ourselves and our children with the skills of metacognition and lifelong learning, we do more than prepare for change—we embrace it. By fostering curiosity, resilience, and a growth mindset, we lay the groundwork for success in a world where adaptability is the ultimate competitive advantage. This shift ensures that as technology advances, humanity not only keeps pace but leads the way.

Conclusion: Harmonizing Human Values with AI Capabilities

Thriving in an AI-driven world demands more than technical proficiency; it requires a deliberate balance between technological advancements and the enduring principles that define our humanity. At its best, AI is not merely a tool for optimization or efficiency—it is a catalyst for innovation, equity, and collaboration. Through an alignment of the technology and its capabilities with human values, we can ensure that it empowers individuals, enhances communities, and addresses global challenges with creativity and purpose.

As we stand on the cusp of a transformative era, the choices we make today will shape the trajectory of tomorrow. By reframing AI as a collaborator rather than a competitor, we unlock its potential to augment human ingenuity, democratize access to knowledge, and bridge divides. Yet, this opportunity comes with responsibility: to integrate AI ethically, to ensure that its benefits are equitably distributed, and to remain vigilant against its misuse.

The path forward is clear. Embrace AI with intention. Cultivate adaptability and curiosity in the face of change. Commit to lifelong learning, and teach the next generation not only how to use AI but how to think critically about its role in their lives. These steps will not only prepare us for the realities of an AI-driven future but also ensure that humanity remains at the heart of this technological revolution.

The challenge is great, but so is the promise. By harmonizing human values with these new capabilities, we can create a future where technology serves as a force for good, advancing both individual potential and collective progress. Let us seize this moment with courage and foresight, shaping a world where humanity and AI thrive together.

About the Author

Dr. James Hutson serves as the lead XR disruptor and head of human-centered AI programming at Lindenwood University, where he pioneers the integration of transformative technologies into education and industry. Holding PhDs in AI and art history, alongside advanced degrees in game design and leadership, Dr. Hutson brings a multidisciplinary lens to his work. He is the founder of the Immersive Arts and Culture Hub and the XR and Gaming Lab, both of which provide cutting-edge resources that empower learners across diverse disciplines to engage with emerging technologies.

Dr. Hutson's academic contributions extend to his role as Department Head of Art History, AI, and Visual Culture at Lindenwood University, where he designs innovative curricula combining traditional scholarship with technological advancements. As editor-in-chief of the *International Journal of Emerging and Disruptive Innovation in Education (iJEDIE),* he facilitates critical dialogue on the future of education. His consulting expertise includes AI integration projects across sectors, helping organizations harness technology to meet real-world challenges.

A prolific researcher, Dr. Hutson has collaborated with over 100 scholars, resulting in even more publications that explore the intersection of AI, education, and culture. His leadership and expertise have earned him accolades from CIO Times and Marquis Who's Who in America, reflecting his transformative impact on education and technology. With certifications in museum studies and organizational change, he embodies adaptability and forward-thinking leadership.

Dr. Hutson's influence reaches a global audience of over 15,000 followers, inspiring professionals and academics alike to embrace AI as a tool for innovation and equity. Through his roles as a mentor, educator, and thought leader, Dr. Hutson continues to shape the future of AI and education.

Email: jhutson@lindenwood.edu

Website: https://www.linkedin.com/in/jameshutsonphd/

CHAPTER 11

FROM AUTOMATION TO ANTICIPATION: REINVENTING CUSTOMER JOURNEYS FOR THE DIGITAL AGE

By Jim Iyoob
Chief Customer Officer, CX Hall of Fame
San Antonio, Texas

Customer experience (CX) has undergone a profound transformation in modern business. This chapter explores the evolution from basic automation to intelligent anticipation, showing how organizations can leverage new technologies while preserving the essential human connection in customer service.

Part One—The Evolution of AI in Customer Experience

The only constant in technology is change, but the principles of excellent customer service remain timeless

---

It started with a simple phone call. Imagine calling customer support in the 1990s. A human agent, flipping through a massive binder of scripts, struggles to find the right answer to your problem. You're placed on hold—again. Frustration builds. Fast-forward to today: you send a quick message via chat, and within seconds, an AI assistant pulls up your history, anticipates your question, and provides a solution—no wait time, no repetition. This transformation from reactive, manual processes to AI-driven, proactive customer engagement marks one of the most significant shifts in business today. But how did we get here? And where are we heading next?

The Early Days: Rule-Based Systems

In the beginning, automated customer service was rigid and formulaic. Think back to the first time you encountered an automated phone system: *"Press 1 for sales, press 2 for support ..."* These early systems were built on simple if-then logic, following strict rules with no room for flexibility or nuance. The limitations were clear: frustrating IVR systems, rigid decision trees, poor natural language understanding, and one-size-fits-all approaches that frequently led to dead ends.

Historical insight: the first automated call distribution systems emerged in the 1960s, but it wasn't until the 1990s that real automation began to take hold in customer service centers.

Today's AI systems represent a quantum leap forward. They follow rules—they learn, adapt, and improve over time. Modern systems can:

- Engage in natural conversations that feel human-like
- Learn from each interaction to improve future responses
- Consider context and customer history in real time
- Make dynamic decisions based on multiple factors

AI has fundamentally changed customer service from a reactive model, where businesses waited for customer issues, to a proactive approach, where AI anticipates and often prevents problems before they arise.

The Rise of Specialized AI Solutions

As AI technology matured, a crucial realization emerged: excellence in customer experience (CX) requires specialized solutions. This insight led to a new generation of AI tools designed specifically for customer interactions. General AI platforms, despite their impressive capabilities, often struggle with:

1. Industry Context

   - Difficulty understanding specialized terminology
   - Limited grasp of industry-specific processes
   - Inability to handle complex, context-dependent queries

2. Customer Sensitivity

   - Challenges managing sensitive information
   - Limited ability to gauge emotional context
   - Struggles with brand voice consistency

The Power of Tailored Solutions

Specialized customer experience AI has emerged as a game-changer, offering:

1. Deep Industry Understanding

   - Mastery of sector-specific language and processes
   - Customized workflows for unique business needs
   - Built-in compliance with industry regulations

2. Enhanced Security and Privacy

   - Purpose-built data protection features
   - Secure handling of sensitive customer information
   - Compliance with industry-specific regulations

3. Seamless Integration

   - Easy connection with existing tools and systems
   - Custom APIs for specific business needs
   - Unified data flow across platforms

Looking forward: as we stand at the cusp of new AI breakthroughs, the evolution continues. The next generation of AI solutions promises an even more sophisticated understanding of customer needs, emotions, and preferences. This transformation in customer experience technology reflects a broader shift in how businesses connect with their customers. It's not just about automating interactions—it's about enhancing them, making them more meaningful, more efficient, and more satisfying for both customers and agents.

Part Two—Building the Modern CX Framework

We've reached a pivotal moment in CX evolution. Despite technological advances, successful organizations recognize that exceptional service requires human connection. Here we explore how leading companies combine technology and humanity to create remarkable customer experiences.

Understanding Modern Conversation Intelligence

At its core, great customer service has always been about understanding—understanding what customers need, how they feel, and how best to help them. Modern conversation intelligence takes this understanding to new levels. Consider how a traditional quality monitoring program might work: a supervisor randomly samples a few calls each week, providing feedback days or weeks after the interaction occurred. Now imagine instead being able to understand every customer conversation as it happens, identifying patterns and opportunities that would otherwise remain hidden.

Historical perspective: this shift to comprehensive analysis marks the most significant advance in contact centers since automatic call distribution emerged in the 1960s.

Modern conversation intelligence provides a level of insight into customer needs and agent performance that was previously impossible. For example, by analyzing every customer interaction, businesses can identify patterns and trends that would be missed by traditional quality monitoring methods.

Empowering Agents for Success

The most effective contact centers recognize that technology should enhance, not replace, human capabilities. Modern support systems serve as an extension of the agent's expertise, providing them with the right information at the right time. Real-time assistance has evolved far beyond simple screen pops. Today's systems analyze conversations as they happen, providing agents with relevant information and suggestions that help them better serve customers. But the key lies in how this assistance is delivered—not as a rigid script to follow, but as a subtle guide that allows agents to maintain natural conversation flow.

Real-world application: a financial services company implemented real-time assistance that provided agents with personalized customer insights. Rather than generic prompts, the system offered context-aware suggestions based on the customer's history and current needs. The result? A 40% improvement in first-contact resolution and, importantly, more confident agents.

The Art and Science of Predictive Analytics

Understanding past behavior helps us serve customers better, but anticipating future needs transforms customer service from reactive to proactive. Effective predictive systems don't just crunch numbers—they provide actionable insights that help organizations better serve their customers.

Finding the Right Balance

When it comes to implementation, one size does not fit all. Organizations must carefully consider their unique needs, customer expectations, and operational requirements. Some key considerations include:

Scalability and Maintenance

Public systems offer the advantage of regular updates and broader knowledge bases, making them attractive for organizations with more general customer service needs. A retail chain successfully implemented a public system for their first-line customer support, significantly improving response times while maintaining high customer satisfaction.

Implementation note: the most successful organizations start small, measuring results carefully before expanding. This approach allows for adjustments based on real-world feedback rather than theoretical benefits.

This framework represents a fundamental shift in how organizations approach customer service. By combining human expertise with modern capabilities, organizations can create experiences that are both more efficient and more satisfying for customers and agents alike.

Part Three—Implementing for Impact: Strategic Integration and Measurement

The gap between vision and execution has derailed many promising CX initiatives. This chapter explores how successful organizations bridge that gap, creating measurable improvements in both customer satisfaction and operational efficiency.

Strategic Integration: Beyond the Basics

The term "omnichannel" has become commonplace in customer service discussions, but truly integrated support remains rare. True integration isn't just about being present on multiple channels—it's about creating a seamless experience regardless of how customers choose to interact.

When customers move from website chat to phone to email, most organizations treat each interaction separately, forcing repetitive explanations. True integration maintains context across channels, creating a single conversation.

Real-world impact: one retail bank implemented channel integration where agents accessed complete customer history regardless of channel. This reduced handle time by 45% while significantly improving satisfaction scores.

The Human Element in Integration

While technology enables integration, success ultimately depends on people. Organizations must consider how integration affects their teams:

1. Training requirements change when agents need to handle multiple channels effectively.

2. Workflow adjustments become necessary as teams adapt to new tools.

3. Performance metrics may need revision to reflect the reality of integrated support.

Implementation note: successful organizations often start with a pilot program, selecting a small team to test integrated support before rolling it out more broadly. This approach allows for adjustments based on real-world experience rather than theoretical planning.

Measuring What Matters

The old adage, "You can't improve what you don't measure," holds particularly true in CX. However, choosing the right metrics—and measuring them effectively—requires careful consideration.

Beyond Traditional Metrics

Traditional contact center metrics like average handle time and calls per hour still have their place, but they tell only part of the story. Modern CX requires a more nuanced approach to measurement:

1. Net promoter score (NPS), while useful, should be considered alongside other indicators. A high NPS doesn't always correlate with operational efficiency or agent satisfaction.

2. Customer effort score provides insight into how easy (or difficult) customers find their service interactions. This metric often proves more valuable than traditional satisfaction scores.

Case study: a telecommunications provider discovered that their focus on reducing average handle time was actually increasing customer effort and repeat calls. By shifting their focus to first contact resolution and customer effort scores, they improved both efficiency and satisfaction.

The Role of Quality Monitoring

Quality monitoring in an integrated environment requires a fresh approach. Traditional quality monitoring programs often focus on individual interactions, but modern CX demands a more holistic view:

1. Consider the complete customer journey across channels.

2. Evaluate the effectiveness of channel transitions.

3. Assess the consistency of information and experience.

Practical example: a healthcare provider implemented journey-based quality monitoring, evaluating complete episodes of care rather than individual interactions. This approach revealed opportunities for improvement that weren't visible when examining single touchpoints in isolation.

Balancing Automation and Human Touch

A common misconception is that AI replaces human agents, when in reality, the most effective customer experiences result from AI *empowering* human expertise. AI handles repetitive, low-complexity queries, freeing agents to focus on high-value, emotionally nuanced interactions where human empathy is irreplaceable.

For example: real-time AI coaching tools analyze conversations and suggest response improvements. Rather than replacing agents, AI serves as an intelligent assistant that guides better interactions—resulting in faster resolutions, improved agent confidence, and more personalized customer experiences.

Creating Effective Handoffs

The moment of transition between automated and human support often proves critical to customer satisfaction. Organizations must carefully design these handoffs to maintain context and momentum:

1. Ensure all relevant information transfers with the customer.
2. Train agents to acknowledge and build upon automated interactions.
3. Create clear criteria for when automation should transition to human support.

Implementation insight: a financial services firm reduced customer frustration by implementing "warm" handoffs from automated to human support. Their system provided agents with a complete interaction history and suggested next steps, making transitions feel natural rather than disruptive.

As we move forward, successful implementation will increasingly depend on:

1. Understanding and adapting to changing customer preferences
2. Maintaining flexibility in systems and processes
3. Continuously measuring and adjusting based on results

Part Four—Private Language Models: Securing the Future of CX

The rise of large language models has transformed customer service capabilities. However, for organizations handling sensitive customer information, public models pose significant privacy risks.

The Shift to Private Models

Public language models, while powerful, send data through external systems. For industries like healthcare, finance, and telecommunications, this approach simply isn't viable. Private models offer a different path—one where customer data remains firmly under organizational control.

Real-world impact: when a major healthcare provider needed to automate patient inquiries, they found public models couldn't meet HIPAA requirements. Their solution? A private model trained specifically for healthcare interactions, keeping patient data completely within their secured environment.

Understanding Narrow Conversational Models

Rather than using a single, general-purpose model, leading organizations deploy multiple narrow models, each handling specific tasks:

1. A claims processing model that understands insurance terminology

2. An appointment scheduling model that manages calendar interactions

3. A medication inquiry model that recognizes pharmaceutical terms

These specialized models excel at their specific tasks while maintaining strict data boundaries.

Task-Specific Excellence

Consider how a financial services company handles mortgage inquiries. Instead of routing sensitive financial information through a general model, they use:

1. A dedicated income verification model

2. A separate credit assessment model

3. A specialized payment processing model

The Privacy Advantage

Private models offer several key advantages for customer data protection:

1. Complete data control: organizations maintain full oversight of how customer information is processed and stored.

2. Specialized training: models learn from relevant, properly anonymized data.

3. Restricted access: information stays within designated boundaries

---

Implementation example: a telecommunications provider deployed narrow models for billing inquiries. These models understand customer account structures and billing terms but can't access or expose personal customer data beyond their specific domain.

Successful organizations typically follow a staged approach:

1. Identify specific use cases.

2. Deploy narrow models for each interaction type.

3. Establish clear data boundaries between models.

4. Monitor and improve while maintaining privacy.

Case study: a regional bank started with a single private model for balance inquiries. As they proved the concept, they added specialized models for loan applications, credit card services, and investment management—each operating within its own secure domain.

As privacy regulations continue to evolve, private language models and narrow AI will become increasingly crucial for customer service

Part Five—The Road Ahead: Future Trends in CX

CX is changing as technology advances and customer expectations evolve. Organizations are finding new ways to balance technological capabilities with the human elements of service. Modern CX is becoming more insightful than simple preference tracking. Organizations now connect critical customer information including:

1. Life events triggering service needs
2. Behavioral patterns signaling changing requirements
3. Opportunities for meaningful proactive engagement

This creates more informed, helpful customer interactions.

Real-world innovation: when a customer asks about investment changes, systems can now identify potential life changes behind the request, helping agents offer more relevant guidance.

The Rise of Emotional Intelligence

While much of today's focus remains on processing efficiency, the next breakthrough in customer service technology centers on emotional intelligence. This isn't about teaching machines to simulate empathy—it's about creating systems that genuinely understand and appropriately respond to customer emotions. Next-generation systems will understand:

1. Subtle conversational nuances
2. Cultural and contextual factors that shape emotions
3. Appropriate levels of response for different situations

The Evolution of the Service Professional

As technology advances, customer service professionals aren't being replaced—they're becoming more crucial than ever. These roles are evolving into sophisticated positions that combine technological fluency with uniquely human capabilities:

1. Experience orchestrators: guiding customers through complex situations

2. Relationship builders: creating and strengthening customer connections

3. Strategic problem solvers: handling nuanced challenges that require human judgment

---

Training innovation: forward-thinking organizations are already developing comprehensive programs that blend technical literacy with advanced soft skills, preparing their teams for this evolved role.

Trust in the Connected Age

As experiences become more connected, privacy and trust are paramount. Leading organizations focus on:

1. Transparent communication: clear data usage explanations, meaningful privacy controls, and consistent respect for choices

2. Strong foundations: scalable infrastructure, regular security audits, and adaptable systems

3. Human-centered development: continuous training, ethical guidelines, and balanced automation

Preparing for Tomorrow

Success in this evolving landscape requires organizations to:

1. Build robust foundations: create flexible systems that can adapt to emerging technologies while maintaining security and privacy.

2. Invest in people: develop teams that combine technical expertise with strong interpersonal skills.

3. Maintain ethical focus: establish clear guidelines for responsible innovation and privacy protection.

4. Foster adaptability: create systems and processes that can evolve with changing customer needs and technological capabilities.

Leadership note: the future of CX is one where AI seamlessly integrates with human expertise to deliver smarter, more empathetic, and proactive service. Organizations that embrace these trends while upholding privacy, trust, and ethical innovation will set the benchmark for excellence in the CX landscape. As AI continues to evolve, the businesses that prioritize both technological advancement and human connection will not only thrive but redefine the very essence of customer relationships.

Conclusion

Embracing a future where technology empowers humanity—at the intersection of technology and human connection, AI's evolution in CX tells a story of transformation. From rigid, automated systems to dynamic solutions that anticipate needs while enriching human service, we've traced the journey from rule-based interactions to the sophisticated, proactive systems defining modern CX.

The future is bright, yet it demands a delicate balance. On one hand, cutting-edge innovations such as private language models, real-time analytics, and predictive systems promise a future where every interaction is more personalized, efficient, and insightful. On the other, the human touch remains irreplaceable— agents who leverage AI to enhance their capabilities, connect more deeply with customers, and resolve complex issues with empathy and expertise.

In this new era, technology is not a replacement for human interaction but a catalyst for elevating it. Companies that successfully blend these strengths will not only meet the evolving demands of their customers but also set new standards in service excellence. As privacy concerns and ethical considerations continue to shape the landscape, responsible innovation will be the cornerstone of sustainable success.

This journey transforms how we perceive and create value in customer relationships. By embracing AI as an enabler of human potential, organizations aren't just reinventing customer journeys— they're crafting a future where each interaction advances a more connected, empathetic, and intelligent world.

About the Author

Jim Iyoob is a thought leader in AI-driven customer experience (CX) transformation. With over 30 years of experience, he leads strategy, marketing, business development, operational excellence, and product innovation across Etech's diverse business lines. Jim has pioneered the integration of artificial intelligence, analytics, and automation in contact centers and digital customer engagement. His expertise in AI-powered chat, predictive analytics, omnichannel CX, and social media engagement has established him as a recognized authority in the BPO and outsourcing industries.

As a trusted industry expert, Jim has been instrumental in helping brands integrate AI and machine learning into their CX ecosystems, ensuring seamless, personalized experiences for customers while improving operational agility.

As an author, Jim has co-written three books with his lifelong friend, mentor, and coach, Matt Rocco, sharing their expertise on leadership, CX, and digital transformation.

LinkedIn: https://www.linkedin.com/in/jimiyoob/

CHAPTER 12

ENTREPRENEURSHIP AND AI: BUILDING TEAMS AND THRIVING THROUGH INNOVATION AND RESILIENCE

By Rafał Janczyk
Founder, AI Pioneer, Team Builder, Triathlete
Łódź, Poland

From my experience building high-tech teams, I've learned that the specific technology doesn't really matter—what matters is the passion for technology. For me, team-building has always been about ultra-focus and dedication. That's how I approached my first AI company, which was born out of my own curiosity and passion for AI.

Back in early 2016, I attended a conference in the Bay Area called AI Frontiers. The keynote speaker was Andrew Ng. At the time, I didn't know who he was, but his talk about AI was so fascinating and convincing that it completely changed my trajectory. I decided to leave my stable, well-paying Silicon Valley job and start from scratch with my own company. The mission was simple: build the best independent AI team and deliver top-notch AI services.

At the time, I was convinced the AI revolution was just around the corner. But there were challenges. Not many people had hands-on experience with practical AI projects, and not many customers were ready to take the plunge into this new space. It was a classic low-demand, low-supply situation, which doesn't exactly scream "great investment opportunity." But I was driven by passion, so I went for it anyway. In this chapter, I'll share my thinking behind building AI teams and how I managed to innovate and grow a team of 100 data scientists and machine learning engineers in a market where skilled talent was hard to come by.

Identify the Mastermind

My first goal was to find a business partner and co-founder with real-world experience and a proven track record in delivering practical AI projects. I decided to build the team and search for talent in my home country, Poland. While Poland has always been strong in software development, I didn't expect to find many data scientists ready to jump on board in 2016—and I was right. To my surprise, LinkedIn showed only three people (yes, just three!) in the entire country who had any hands-on, non-academic AI experience at the time. And this was in a tech-savvy, innovative country of 38 million people!

Luckily, two of them agreed to meet with me, and one of them—Łukasz—was not only incredibly skilled but also genuinely passionate about the idea of building an AI services company. He decided to join me on this journey. Łukasz, I'll always say it: I was lucky to find you.

The Source of Talent

As there were no ready-made, trained data scientists on the market, we had to find a way to recruit and train people who would be able to absorb the knowledge most efficiently. We identified a few sources of data science apprentices that we could take with us on the AI journey: computer scientists, econometricians, statisticians, mathematicians, and physicists. The last category could come with an anecdote. One of the early team members was a quantum physics PhD. When I asked him at the interview why he had decided to exchange his field of science for data science, his answer surprised me and made me laugh. "I found there is nothing more challenging than astrophysics, so I decided to step down and focus on something simpler and more practical—AI."

The Training Program

The ideal time to plant a tree was 30 years ago, but the second-best time is now. Keeping this in mind, and with the head of data science and several apprentices already in place, we began assembling our first team. Since we recruited a diverse group with no prior experience in AI, we recognized the need to establish a training program to help them gain the necessary skills in this emerging field. As a result, we developed a custom AI academy program. Although it started with a rough outline of the skills needed for typical data scientists, it has now become a solid foundation for nurturing and growing our talent.

Team Culture

When the first team came together, I was surprised by how different the culture felt compared to the world of software developers I was used to. As a computer scientist, I expected them to be part of my tribe— sharing the same approach to problem-solving, tackling challenges, and daily work life. But I was wrong. The over 100 data scientists I've had the luck to work with over the years felt like a completely new species.

Now, after more than eight years of observing them, I'd say programming is like craftsmanship while data science is more like artistry. This will probably evolve as more practical AI projects are delivered and as AI product development matures. But back then, my world of nerdy developers collided with a world of *uber-nerds*—people with even more passion, excitement, and their own ecosystem of jokes, memes, and quirks.

When it came to fostering the team culture, I realized the best thing I could do was not to interfere. The process of self-creation was amazing to watch. My role was mostly to stick to the founding vision: *we want to build the best independent AI team in the industry.* That was the only direction and boundary I provided—this incredible group of people took care of the rest.

What really helped the culture grow naturally was the team's eagerness to share knowledge. Most of the team joined to learn data science, but they were just as excited to teach what they'd learned to each other and to newcomers. This "student today, teacher tomorrow" mindset came from the void around us. At the time, there was only one other company like ours in the entire country, but we knew where to find the knowledge we needed. That sense of being pioneers really opened up communication and created a culture of sharing in a field that was evolving at lightning speed.

The Magic of the First Team

I love starting over and every time I do it, I've noticed something magical about the first team. Those first 20 people are always special. They're true pioneers—willing to go the extra mile, unfazed by a scrappy office or the lack of fancy perks like jasmine tea in the kitchen. These are the kind of people who join a new company because they're looking for freedom and the chance to be part of the act of creation.

What's fascinating is watching their careers grow within the organization. They tend to develop a strong sense of attachment to the company they helped build, and they often naturally step into leadership roles, becoming the first wave of management. It's like they're not just employees; they're co-creators.

Don't Look for Unicorns—Build a Diverse Team

You can probably picture the perfect unicorn data scientist: someone with ten years of experience in applied AI, 30 successful projects under their belt, great leadership skills, excellent customer-facing abilities, and the talent to understand data and present insights to both technical teams and senior executives. These people do exist, and you should absolutely be open to hiring or developing that kind of talent. But let's be real—you're not going to find or grow many of them. Instead, my advice is to focus on building a mixed team with a variety of skills. Here's what you'll likely need:

- *Researchers (up to 10% of your team)*: these are your PhDs—former academics who excel at finding the best solutions, algorithms, and models quickly.

- *Applied data scientists (up to 40% of your team)*: these team members should be proficient in Python or another scripting language, able to efficiently work with limited data sets, and focused on delivering project goals.

- *Machine learning engineers and software developers (up to 50% of your team)*: these are the people who can take a promising model and make it work in the real world. They're skilled programmers who understand that a machine learning model doesn't exist in a vacuum—it needs to run on hardware with limited resources.

By balancing these roles, you'll create a team that's not just chasing unicorns but actually getting things done.

First Projects: Hands-On Experience

Once you've built your team, it's important to keep their curious and passionate minds engaged and learning. There are a few tried-and-true strategies for this. One is to pair new team members with experienced ones, so knowledge can flow naturally. Another is to fund an internal project that, while not a top priority for the organization, can still deliver a big boost in motivation and morale if it succeeds.

For us, the first project was all about decrypting Enigma machine code using artificial intelligence. It was a perfect fit since my first AI company was called Enigma Pattern. Not only did this quirky project give us some great publicity, but it also gave our brand-new team a chance to cut their teeth and gain some real experience.

Project Management for AI Projects Matters

In my experience, data science projects are very different from regular software development projects, especially during the initial research phase. Managing a nondeterministic environment—where outcomes are open to interpretation and might not align with the original goals or assumptions—requires a different skill set than leading a software team.

That's why it's crucial to find someone who understands how to work with data scientists, data sets, machine learning (ML) ops, and the unique challenges of AI projects. This person needs to know that the goal isn't just a working app, but a machine learning model that has to meet specific performance metrics, like precision and recall and solve the business problem of the customer.

Stay Innovative and Keep an Eye on New Technologie

Being a pioneer paid off—our company quickly became the number one AI employer in Central Europe. We promised our employees they could focus purely on AI and gave them access to a wide range of multidisciplinary data science applications. Think about it: a super curious mind stuck in finance, automotive, or any other single industry for five years without the chance to explore new areas. For us, offering diverse projects was key to attracting top talent. It was our way of standing out in the global battle for tech skills.

We also realized early on that AI evolves at lightning speed—like a Shinkansen bullet train. That's why we always kept a close eye on the latest trends and launched internal projects to test and validate emerging technologies. One of my favorite tools for tracking this is Gartner's Hype Cycle Curve. If you take a look at curves published in 2019 and 2024 you can see for instance how Computer Vision that was no longer a hot topic and no longer in a peak of hype in 2019 turned into mass production in 2024. It's simply fascinating to see how much progress some technologies have made in just five years.

Resilience in Finding Solutions

When we first started, we didn't have much choice but to take on most projects, even when some of our customers' ideas were really ambitious. I remember having constant debates with our head of data science about where to draw the line on risk—when to say yes to a project and when to walk away. To manage this, we came up with a clear framework for project maturity, breaking it down into stages:

1. *Research:* we'd start by digging into the latest scientific papers to see if the customer's idea was even feasible according to academia. To move to the next stage, we needed to confirm that there was an existing algorithm we could use as the foundation for a solution.

2. *Proof of Concept (PoC):* this was a short project, no longer than three months, using a limited data set. The goal was to prove that the customer's idea could work with their data and produce a machine learning model that delivered the expected results. If we hit the right precision and recall metrics, and the effort and cost fit the customer's budget, we'd move to the next stage.

3. *MVP (minimum viable product):* this was the first functional version of the product, designed to test the core idea with real users.

4. *Full product with support and maintenance*: once the MVP was validated, we'd build out the full product and provide ongoing support.

This approach helped us stay focused, manage risk, and deliver results without overpromising.

Consider Outsourcing If You Don't Plan to Build a Big Team

If you're not planning to build a large team or if you're creating a software product that doesn't need rapid scaling, outsourcing might be your best bet. Here's the unspoken truth about data science projects: they're usually short-term and don't require massive teams.

In my journey with Enigma Pattern and later Digica, I've been lucky enough to oversee more than 300 AI projects. The biggest one—an intensive research project on synthetic data in both RGB and infrared spaces—required just seven people for nine months. That's why I'd encourage you to reach out to established data science service providers or AI consultancies. It saves you the hassle of recruiting rare talent or taking a risk on people with unproven track records. Yes, you'll pay top dollar, but think of it as insurance for your journey into the unknown.

Things to Avoid

After eight years and 300 projects, I've learned a lot about what *not* to do when building and managing an AI team. Here's my advice:

1. *Data Science Isn't Software Development*

Don't treat the research phase like a typical software project. Plan for shorter sprints to allow flexibility and frequent pivots—because in data science, you'll need them to avoid falling into a dead end of endless research.

2. *But It's Still Software Development*

While research is different, you still need to plan for a final product, not just a proof of concept (PoC). Think about the target platform early on—will this run in the cloud or on an edge device? Get one of your software architects involved from the start.

3. *The Curse of a Successful PoC*

It's relatively easy to prove an idea works in a PoC, but turning that into a full product is much harder. The data set might be sufficient, but the computing power on the edge device might not be—or vice versa.

4. *The Approval Process—Plan for Delays*

Most organizations pay extra attention to anything AI-related, which can slow things down. Plus, inexperience in the product acceptance process can lead to delays and require your senior data scientists to spend extra time hand-holding customers.

5. *Defense Projects Come with Constraints*

If you're working on defense projects, be prepared for some team members to opt out due to ethical concerns. It's something to keep in mind when assembling your team.

6. *Remote Work Doesn't Work for AI Development*

If you're serious about building AI solutions, remote work won't cut it. The magic happens in the office—in the cafeteria, in the hallways, during casual conversations. If you're not ready to bring people together in person, consider outsourcing. Otherwise, you'll struggle to create that tight-knit, collaborative culture that feels like you're part of the Manhattan Project.

Final Thoughts

If you share my passion for AI and the incredible advancements it can bring, here's my message to you: the AI train is still at the station, and there's plenty of room for you to hop on and join the journey. Whether you're a software engineer, a project manager, a general manager, or the CEO of a traditional company, there's a place for you in this space.

I can promise you this: it's an exciting ride, and opportunities like this don't come around often. The last time we saw something this groundbreaking was during the dot-com boom in 2000, and before that, it was the rise of microcomputers in the 1980s. Who knows how long we'll have to wait for the next big thing—maybe another 20 years? So don't wait. Come and be part of this revolution.

About the Author

Rafał Janczyk is a serial entrepreneur and passionate team builder who has emerged as an AI pioneer. Inspired by his decade spent in Silicon Valley, he decided to pursue his passion for AI and founded a company that has become the largest independent AI services provider in Europe. With eight years of experience delivering practical AI solutions and over 300 projects completed, Rafał is a leader in the field.

In his personal life, Rafał enjoys spending his free time with his family and competing in triathlons, with the aim of participating in an Ironman competition each year.

Email: rafal.janczyk@gmail.com
LinkedIn: https://www.linkedin.com/in/rafal-janczyk

CHAPTER 13

THE DAWN OF SYNERGY— HUMAN AND AI

By Susmitha Jella
Executive Leader, Global Awards, Speaker, AI & Innovation
Melbourne, Australia

> *It's not about what it is, it's about what it can become.*
> —Dr. Seuss

The Arrival of Artificial Intelligence—The Next 100 Years

We stand at the precipice of a new epoch where artificial intelligence (AI) is not just a tool but a cornerstone of our existence. Just as the Industrial Revolution transformed human life, so too has AI become the foundation that unlocks the future's limitless potential. It's more than a technological marvel; it's the very key to our capacity to function and innovate. From diagnosing diseases to optimising every corner of our daily lives, AI has marked its presence.

The next 100 years will be defined by the interplay between human evolution and the rapid advancement of AI. This chapter explores the core values driving our need for AI, the ethical dilemmas we can face, and the potential consequences of our actions—or inactions. It also examines the delicate balance between leveraging AI for social good and the risks of creating systems that could outpace our control. We are experiencing an era of information overload, amassing vast amounts of data. It is essential to discern what is valuable and significant to achieve relevant and meaningful AI outcomes. We stand at the threshold of a transformative era, and we should ask ourselves: what is our purpose with AI? Where do we draw the line? What kind of future do we want to create with AI?

Scratching the Surface—Core Values and the Purpose of AI

AI has existed for decades, but what makes it truly revolutionary today is its unprecedented accessibility—now literally at your fingertips. This shift has transformed once-fictional concepts into tangible realities, reshaping how we live, work, and interact with the world.

As we navigate this digital age, it becomes evident that we are merely scratching the surface of AI's possibilities. The promise it holds is vast, and its full spectrum is yet to be explored. Each day brings forth new discoveries, and with them, we realize the boundless potential that AI harbors, poised to elevate our collective human experience. Every day a new potential of AI is discovered.

At the heart of humanity's drive to develop AI lies a set of core values: the pursuit of knowledge, the desire to improve quality of life, and the need to solve complex global challenges. AI has the potential to amplify these values, enabling us to tackle problems that have long seemed insurmountable. From curing diseases to addressing climate change, AI offers tools that can augment human capabilities and accelerate progress. But is it all about the money? While economic incentives undeniably fuel AI development, reducing its purpose to mere profit would be a grave oversimplification. The true purpose of

AI should align with humanity's broader aspirations: to create a more equitable, sustainable, and enlightened world. This requires us to move beyond short-term gains and consider the long-term implications of our creations.

The Tapestry of Technology and Imagination

For many, the idea of AI isn't a new concept. Generations have grown up watching science fiction narratives where technology and AI power the universe. Iconic stories like *Star Trek* have etched the notion of a possible harmonious coexistence between humans and intelligent machines into our cultural fabric. These fictional tales, often dismissed as mere fantasy, now serve as blueprints for our future. The reality of a human AI is closer than we think. These tales have created the art of the possible.

Digital Realms and Futuristic Visions: Exploring Cyberspace, Cyberpunk, and Superhero AI

As we take a glimpse into the future, we find ourselves at crossroads: are we on the brink of a world dominated by superhero AI, a force for unparalleled good, or superhuman AI, which might rival us in our capabilities? The line between AI and humanity blurs, inviting us to ponder the roles these entities will play in shaping our destinies.

With the evolution of AI, over the next century, the notion of setting up homes on Mars and establishing new human colonies doesn't seem far-fetched. With AI driving advancements in space exploration, data management, and resource optimization, the dream of interplanetary life inches closer to reality. AI could be the catalyst that propels humanity to new celestial frontiers.

Countless stories have depicted AI as both a formidable threat and a benefactor, painting a picture of dystopian takeovers and utopian symbioses. The fear of AI usurping human authority is juxtaposed with visions of AI enhancing human life in unimaginable ways. Amidst this duality, we must seek to understand where we truly stand. Let's ask ourselves—*what do we want to achieve with the revolution of AI?* Will we hurtle towards an era of cyberpunk dystopia, where AI's misuse leads to societal breakdown? Or will we forge a future that harnesses AI's strengths without compromising human values? The human element remains crucial in this 100-year fantasy, as we envision a world that balances technological advancement with ethical considerations.

Navigating the Future: Servant or Sovereign, and Regulation

As AI continues to evolve, questions arise about its ultimate role. Will AI serve humanity, augmenting our capabilities and enhancing our lives? Or will we witness the rise of sentient AI, entities that can toggle emotions and values as easily as flicking a switch? The future may hold AI that feels, thinks, and even rivals human creativity.

"With great power comes great responsibility"—as popularised by the Superhero Marvel Universe. The potential for AI to go rogue poses significant risks. A rogue AI, unchecked and unregulated, will make the dystopian narratives we feared a reality. AI could surpass human control and lead to a fragmented society. Ensuring robust safeguards and ethical frameworks will be paramount to prevent such eventualities.

What happens if we don't act and we fail to establish robust standards and ethical frameworks for AI? We risk creating a future where technology outpaces our ability to manage it. Without proper regulation, AI's unchecked growth could unleash unintended consequences. Proactive regulation and ethical guidelines will be vital in steering AI's trajectory.

One of the most profound risks is the potential stagnation—or even regression—of human intellect. We've seen this happen with social media—it is a double-edged sword, capable of driving positive change and fostering global connectivity, yet equally prone to perpetuating harm, as evidenced by its role in spreading misinformation, amplifying polarization, and eroding mental health over the past 15 years.

As AI systems become increasingly capable of performing tasks that once required human thought, there is a danger that we may become overly reliant on them, leading to a decline in critical thinking, creativity, and problem-solving skills.

The unchecked evolution of AI could lead to a scenario where humanity's role in shaping the future is diminished. If AI systems are allowed to evolve without human oversight, they could develop goals and behaviours that are misaligned with our values. This misalignment could result in unintended consequences, from the erosion of privacy and autonomy to the exacerbation of social inequalities.

Socio-Economic Benefits, Standards, Ethical Use, and Challenges

To avoid pitfalls, we must prioritize the development of standards, processes, and ethical guidelines for AI. This includes establishing transparency and explainability in AI decision-making, ensuring accountability for AI-driven outcomes, and promoting fairness in the distribution of AI benefits. Ethical AI development also requires us to consider the impact on vulnerable populations and to guard against the misuse of AI for surveillance, manipulation, or harm.

One of the key challenges is determining where to draw the line. While AI can be a powerful tool for social good—such as analyzing vast amounts of data to predict natural disasters or identify patterns in disease outbreaks—it can also be used in ways that undermine human dignity and autonomy. For example, the use of AI in surveillance and predictive policing raises concerns about privacy and civil liberties. Similarly, the deployment of autonomous weapons systems poses ethical and moral dilemmas that demand careful consideration

AI's integration into various sectors promises profound socio-economic benefits. It can drive unprecedented efficiency, reduce costs, and improve overall quality of life. However, this progress comes with its challenges—economic disparities, job displacement, and the need for continuous learning and adaptation.

The development and deployment of AI also has significant implications for the environment and natural resources. Training advanced AI models requires vast amounts of computational power, which in turn consumes substantial energy and contributes to carbon emissions. As AI becomes more integrated into our lives, we must consider its environmental footprint and explore ways to make AI development more sustainable—and question *when* AI should be used in our daily lives. Additionally, the extraction of rare earth metals and other resources needed for AI hardware raises concerns about environmental degradation and resource depletion. To address these challenges, we must prioritize the development of energy-efficient AI systems and invest in sustainable technologies that minimize the environmental impact of AI.

Resilience and Vulnerabilities

Is AI resilient? Like any technology, AI is not immune to vulnerabilities. These vulnerabilities can be exploited by malicious actors to manipulate AI systems, leading to outcomes that undermine their intended purpose. For example, adversarial attacks on AI algorithms can cause them to make incorrect decisions, with potentially devastating consequences in critical areas such as healthcare, transportation, and national security.

The risks associated with AI vulnerabilities are compounded by the potential for misuse. In the wrong hands, AI could be used to spread misinformation, manipulate public opinion, or even wage cyber warfare. The lines between beneficial and harmful uses of AI can become so murky that humanity may face a stark choice: to either regulate and control AI or risk its destruction.

The Human-AI Nexus: Functionality and Identity

As we embark on the journey of augmentation, we should take a philosophical approach to the human and AI conundrum. A human being is understood to be made up of approximately 60 billion neurons, forming a complex network that gives rise to consciousness, emotion, and creativity. As AI systems become more sophisticated, we must ask ourselves: do we want AI to function like a human? While replicating human cognition may seem like a logical goal, it raises profound questions about identity, agency, and the nature of intelligence.

Creating AI that mimics human thought processes could blur the line between human and machine, challenging our understanding of what it means to be human. It could also lead to ethical dilemmas, such as whether AI systems should be granted rights or treated as moral agents. Instead of striving to replicate human intelligence, we should focus on developing AI that complements and enhances human capabilities. To thrive in a future dominated by AI, augmenting human intellect and artificial intelligence is essential. Understanding when to leverage AI versus human intuition and creativity will be crucial. A delicate balance must be struck to achieve outcomes that benefit society.

The next century could witness the rise of a harmonious coexistence between AI, robots, nature, and humans. A symbiotic relationship where AI enhances human potential, bots perform mundane tasks, and humanity focuses on creativity, whilst compassion is the goal of our existence.

The next 100 years will be shaped by the choices we make today. If we approach AI development with a focus on collaboration, transparency, and ethical responsibility, we can create a future where AI serves as a powerful tool for human advancement. However, if we fail to act responsibly, we risk creating a future marked by conflict and instability. The unchecked evolution of AI could lead to a world where humanity is overshadowed by its own creations, struggling to maintain control over systems that have grown beyond our understanding. In such a scenario, the destruction of AI—or even of the natural world itself—could become a grim possibility.

As we grow in the AI revolution, human oversight should remain the ultimate authority in decision-making, ensuring that ethical considerations, critical judgment, and accountability guide the outcomes of AI-driven processes. Responsible AI should serve as a foundational advisory framework, establishing essential guardrails to ensure the ethical development, deployment, and management of AI systems. As AI models and their applications continue to expand, this framework must prioritise transparency, accountability, and fairness, safeguarding against misuse and unintended consequences. By acting as a protective layer, responsible AI should not only guide the responsible use of AI but also ensure the secure and ethical handling of sensitive information, fostering trust and confidence in AI technologies across industries and society.

Conclusion: A Vision for Tomorrow

As we embark on this journey into the next 100 years, we must remain vigilant, hopeful, innovative, guided by our core values with a clear sense of purpose. AI, if wielded wisely, holds the promise of a brighter, more connected, and prosperous world where humans and machines thrive together. The next century offers unprecedented opportunities to harness AI for the benefit of humanity, but it also presents profound challenges that demand careful consideration and action. By building standards, driving ethical use, and fostering collaboration, we can ensure that AI evolves in a way that aligns with our aspirations and preserves the beauty and diversity of the natural world.

The question of where to draw the line is not one that can be answered easily, but it is one that we should confront with courage and foresight. The future of humanity and AI is intertwined, and the choices we make today will shape the course of our shared evolution for generations to come. Let us choose wisely.

About the Author

Susmitha Jella is a visionary leader at the forefront of the data and AI revolution. She transforms data and AI challenges into enlightening journeys, deriving actionable insights that drive smarter decisions. Su's strategic vision and leadership have revolutionised decision-making processes, operational efficiencies, and growth trajectories in top-tier organisations. Her approach uniquely integrates human-centric perspectives with data and AI strategies. As a modern leader, she champions continuous learning, coaching, and agility, empowering thousands of professionals through innovative workshops and discussions. She shares cutting-edge insights on data and AI governance, strategy and ethical leadership, inspiring a new generation of leaders with data proficiency with human-centric values.

Recognised as one of the Global Top 50 Innovators in Data and Analytics, Su is a winner in the Women in AI (Asia Pacific) award and is also listed among the Top 25 Leaders in Australia.

Email: su@dewdropdigital.com.au

Website: www.dewdropdigital.com.au

CHAPTER 14

AI AND ENERGY: LEADING INNOVATION IN TECHNOLOGY SOLUTIONS FOR A SMARTER FUTURE

By Kent B. Landrum
Partner; Process & Technology Leader
San Antonio, Texas

> *AI is the new electricity.*
> —Andrew Ng

The energy industry is the backbone of modern society, powering homes, industries, and economies while enabling technological advancements, infrastructure development, and overall quality of life. In a rapidly evolving world, artificial intelligence (AI) is transforming how we approach energy production, distribution, and consumption. From optimizing resource use to enhancing the sustainability of energy sources, AI stands at the forefront of creating a more innovative, more efficient energy future.

The energy industry is no stranger to complex, data-intensive challenges, and AI offers a potent toolkit to navigate these intricacies. By unlocking the potential of big data and machine learning, AI is steering the energy sector toward new, efficient paradigms that promise economic and environmental benefits on a global scale. Understanding the past, fostering innovation, and ensuring ethical, sound leadership—whether from innovators, executives, or policymakers—will be critical to unlocking AI's full potential in the energy industry, creating an environment that accelerates progress while safeguarding reliability, sustainability, and long-term value for all stakeholders.

AI and the Energy Sector: A Synergistic Partnership

The marriage of AI and energy is no accident. The industry faces challenges particularly well-suited for AI solutions—fluctuating demand, infrastructure strain, and the growing emphasis on renewable resources all demand real-time insights and rapid decision-making. AI technologies, particularly machine learning, neural networks, and predictive analytics, reshape energy models from top to bottom. These tools enable real-time adjustments, cost savings, and greater efficiency by predicting demand, managing supply, and optimizing resources.

A Brief History of AI in the Energy Industry

AI has been evolving for decades, steadily transforming industries—including energy—by enhancing efficiency, cost reduction, and sustainability. While the energy sector is traditionally associated with wildcatters and roughnecks, AI's integration has progressively reshaped decision-making, data analysis, and operational processes. AI's earliest applications in energy emerged in the 1970s, focusing on expert systems designed to replicate human decision-making in areas like seismic data interpretation and extraction optimization.

A notable example is Schlumberger's Dipmeter Advisor, developed with MIT researchers, which helped interpret geological formations using a rule-based system and pattern recognition. However, limited computing power and data availability restricted these early efforts. At the same time, AI began impacting load forecasting, a critical function in power system management. The development of the backpropagation algorithm by Paul Werbos in the mid-1970s enabled the training of artificial neural networks (ANNs), laying the foundation for more sophisticated energy forecasting models. By the 1980s, advances in digital electronics and distributed computing allowed for broader AI applications, reinforcing its role in predictive modeling and decision support.

The 1990s and 2000s marked the expansion of machine learning in energy applications, despite coinciding with the second AI winter, a period of reduced funding due to unmet expectations. As computing power grew and machine learning algorithms advanced, AI became more practical for handling large data sets in reservoir modeling, extraction optimization, and maintenance scheduling. A breakthrough was using ANNs for reservoir characterization and production forecasting, enabling more accurate subsurface modeling and production predictions. AI also enhanced condition-based maintenance, first developed in the 1940s, by integrating predictive analytics to anticipate equipment failures, setting the stage for today's machine learning-driven predictive maintenance systems.

These early AI applications—expert systems, load forecasting models, and ANNs—pioneered modern innovations in predictive analytics, grid optimization, and intelligent automation. Even through the challenges of the AI winters, the persistence of visionary leaders and researchers was crucial in advancing AI's role in energy. Their resilience and commitment to innovation ensured that AI would evolve into a transformative force, driving today's breakthroughs in efficiency, sustainability, and intelligent decision-making.

Energy Consumption Forecasting and Demand Response

Forecasting energy demand has always been a challenging task for utilities. Demand forecasting in the electric power sector involves predicting future electricity consumption using historical data, weather patterns, economic indicators, and real-time grid information. Accurate forecasts enable utilities to optimize generation, reduce operational costs, and ensure grid reliability. Traditional methods were often limited in accuracy, relying on historical data and limited predictive factors. However, AI can process enormous datasets, analyzing patterns and trends that reveal more about potential future energy needs. By incorporating weather patterns, time of day, and consumer behavior, AI-powered forecasting tools can predict energy demand with unprecedented accuracy. This capacity for demand forecasting is particularly critical for renewable energy integration. Renewable sources like wind and solar are inherently variable— producing power only when the wind blows or the sun shines. With AI-driven forecasting, grid operators can better plan for these fluctuations. Improved forecasting ensures steady supply even as renewables take a larger share of the energy mix, preventing energy shortages and reducing reliance on carbon-emitting backup sources like coal or natural gas.

Demand response is another area where AI is making a significant impact. Demand response programs typically incentivize consumers to reduce or shift their electricity use during peak times, relieving grid stress. Many are likely already familiar with demand response through smart thermostat programs, where homeowners receive incentives to allow minor adjustments to their air conditioning during peak demand periods. This same concept is also applied in large commercial and industrial contexts. A manufacturing plant can temporarily scale down non-essential production processes or shift energy-intensive operations to off-peak hours in response to price signals or direct requests from the grid operator, helping to balance electricity demand. With AI, utilities can automatically analyze consumption patterns and send signals to smart devices—like thermostats, water heaters, or electric vehicle chargers—optimizing energy usage in real time. By doing so, AI helps balance the grid, lowers costs, and reduces carbon footprints, better aligning economic and environmental objectives.

Enhancing Renewable Energy Integration and Efficiency with AI

The shift toward renewables has become a central priority in combating climate change. However, integrating these sources into the traditional energy grid presents noteworthy challenges due to their intermittent nature, the need for advanced infrastructure, real-time balancing of supply and demand, and the complexity of managing decentralized generation. AI's role in optimizing the efficiency and reliability of these renewable energy assets has the potential to be transformative. In wind energy, machine learning algorithms analyze extensive weather data, including wind speed, direction, and temperature, to accurately predict wind patterns. These predictions enable real-time adjustments to turbine settings, such as blade angle and rotational speed, optimizing performance, maximizing energy output, and reducing wear and tear on equipment. In solar power, AI-driven analytics can monitor panel efficiency and suggest positioning adjustments, maximizing the energy captured from each installation.

Moreover, AI is invaluable in developing "smart" energy grids that dynamically manage electricity flows using real-time data. Smart grids can balance supply and demand more effectively by distributing renewable energy where it is needed most. AI's role in managing the complexities of these grids, handling fluctuating inputs, and maintaining service reliability is essential to future energy sustainability.

AI and Smart Grids: Toward a Decentralized Energy Future

The traditional energy model—centralized electricity production at large power plants that is then transmitted and distributed to consumers—faces new challenges in an era of distributed energy resources (DERs), such as rooftop solar panels, home batteries, and electric vehicles (EVs). Electric vehicle owners can optimize their energy usage by discharging their EV battery during peak demand hours, such as in the evening when solar generation declines, to support the grid and potentially earn incentives. Conversely, intelligent overnight charging—taking advantage of lower time-of-use (TOU) rates—ensures the vehicle is fully charged for the morning commute while minimizing costs and alleviating grid stress during high-demand periods. As more consumers produce and store their energy, the grid must adapt to two-way energy flows, where consumers can feed power back into the system.

With AI-driven energy management systems, smart grids can allocate energy more efficiently, especially during peak demand. These systems can monitor the status of DERs in real time and integrate them with large-scale utilities, helping to stabilize the grid and provide backup power when needed. DERs can be aggregated and managed through advanced software to function as a single, flexible power source referred to as a virtual power plant (VPP) composed of a network of solar panels, wind turbines, battery storage systems, and demand response capacity for use in grid stabilization and energy market participation. AI facilitates a decentralized model where energy can flow more freely and is less reliant on large, centralized power plants. This shift could lead to a more resilient, sustainable grid that better integrates renewables and reduces greenhouse gas emissions.

AI and Grid Reliability in a Changing Climate

As climate change intensifies the frequency and severity of extreme weather events, ensuring grid reliability has become a critical challenge for energy providers. AI-driven technologies are emerging as powerful tools to enhance asset maintenance, outage prediction, and system resiliency, helping utilities anticipate disruptions and proactively protect infrastructure.

AI transforms asset maintenance and optimization across the power grid, from power plants to transmission lines, poles, and underground cables. Predictive maintenance powered by AI analyzes vast datasets from equipment sensors to detect subtle signs of wear and failure before they lead to costly outages. This approach extends equipment life, minimizes emergency repairs, and reduces operational risks. AI-powered digital twins also improve grid resiliency by creating virtual replicas of power systems that can simulate the effects of severe weather, equipment failures, and operational changes. These digital models allow utilities to test mitigation strategies without real-world risks, accelerating innovation in climate resilience planning.

Beyond individual assets, AI can optimize undergrounding strategies by identifying critical locations—such as root nodes of distribution networks and high-risk areas with a history of failure—where targeted undergrounding cables can dramatically improve reliability. Utilities can create a more resilient system, reducing outage risks from falling trees and mitigating wildfire threats by combining targeted undergrounding with AI-driven vegetation management. By analyzing rainfall patterns, temperature trends, and growth projections, AI systems can predict where trees and vegetation will pose the most significant risk to overhead power lines. Utilities can then proactively dispatch crews to trim trees in high-risk areas.

AI also plays a vital role in forecasting climate-related risks by using weather and storm prediction models to anticipate the impact of hurricanes, wildfires, ice storms, and other extreme events. Risk analysis can better predict power line damage, flooding risks, and infrastructure vulnerabilities, enabling utilities to pre-position crews, reinforce critical assets, and deploy resources efficiently for rapid restoration. For example, AI can forecast wildfire risks by correlating weather conditions, drought severity, and vegetation density, allowing utilities to implement preemptive de-energization strategies or vegetation-clearing efforts in high-risk areas. Similar analytics can predict freeze-offs in natural gas production, hail damage to solar farms, and equipment stress from extreme heat, allowing operators to take preventive actions before failures occur.

AI and Sustainability in Energy Production

The global energy sector is a leading contributor to greenhouse gas emissions. As a result, many companies are embracing AI to advance sustainability. AI-driven dynamic generation scheduling enhances power plants' operation by continuously analyzing real-time fuel costs, grid demand, and renewable energy availability. These intelligent systems optimize dispatch decisions, ensuring that the cheapest, cleanest, and most efficient fuel mix is used at any moment. By seamlessly balancing traditional and renewable energy sources, AI helps reduce reliance on fossil fuels while maintaining grid stability. Beyond scheduling, AI optimizes emissions and efficiency in thermal power plants. By monitoring fuel combustion processes in real time, parameters can be adjusted to maximize energy output while minimizing fuel waste and emissions, leading to lower carbon footprints, reduced operational costs, and improved compliance with emissions regulations.

AI can also play a role in expanding the circular economy in the energy sector by helping companies reduce waste, extend the life cycle of materials, and increase resource reuse. One tangible example is the recycling of wind turbine blades and solar panels, which have historically been challenging due to the complexity of composite materials. AI-powered material science innovations are helping identify new ways to break down, repurpose, and recycle these materials into new energy infrastructure. For instance, AI algorithms optimize chemical processes for breaking down fiberglass from decommissioned wind turbines, turning them into raw materials for new construction or energy storage solutions. Similarly, AI-driven predictive analytics help optimize the lifespan of solar panels by analyzing degradation patterns and recommending proactive maintenance or repurposing strategies. Some companies even use AI to sort and refurbish used solar cells, giving them a second life instead of sending them to landfills.

AI Challenges and Ethical Considerations in the Energy Sector

While AI holds immense promise for transforming the energy sector, its deployment is not without challenges. From data privacy and algorithmic bias to grid security risks and regulatory hurdles, energy companies must navigate these complexities to ensure AI-driven innovations are both practical and ethical. AI systems in the energy sector rely on vast amounts of consumer and operational data, raising significant privacy concerns. Energy consumption patterns, for example, can reveal sensitive details about an individual's daily habits, lifestyle, and even security vulnerabilities. This data could be misused, breached, or exploited for unintended purposes without proper safeguards. To address these risks, energy companies must implement rigorous cybersecurity protocols, robust data governance frameworks, and transparent policies on ethical data usage. Beyond privacy, AI also introduces risks of algorithmic bias, which can lead to unfair energy pricing and access outcomes. If AI-based pricing models prioritize profits over equitable access, marginalized or low-income communities could face higher energy costs, limited access to renewable programs, or exclusion from energy-efficiency incentives. Ensuring fair, transparent, and explainable AI algorithms is critical to preventing such disparities and maintaining public trust.

As intelligent AI-driven systems become increasingly integrated with the operational technology (OT) networks that control energy infrastructure, they introduce new cybersecurity vulnerabilities. AI-enabled automation, predictive analytics, and distributed energy management depend on real-time data exchange across interconnected networks, making them potential targets for cyber threats. A successful attack on AI-powered grid management systems could lead to power disruptions, economic losses, or widespread blackouts. Organizations like the Electric Power Research Institute (EPRI) have highlighted the urgent need for stronger AI cybersecurity measures, including robust threat detection systems, AI-driven anomaly detection, and proactive defense mechanisms. As the energy sector becomes more digitally connected, companies must balance AI innovation with enhanced cybersecurity strategies to prevent disruptions and protect critical infrastructure.

The Road Ahead: AI and the Future of Energy Innovation

As we look to the future, the integration of AI in energy systems will continue to accelerate, driving innovation and creating more intelligent, sustainable solutions. AI's role will likely expand into nuclear fusion, advanced battery storage, and grid resilience against climate-induced events in the coming years. With the right strategies, AI can help the energy sector achieve its dual goals of economic efficiency and environmental stewardship. AI's transformative power is already evident in areas ranging from demand forecasting to renewable integration, market operations, and sustainability initiatives. Ironically, while AI is being leveraged to enhance energy efficiency and sustainability, its energy consumption is a growing concern. Training complex machine learning models requires significant computational power, leading to high electricity demand—sometimes contradicting the sector's broader sustainability goals. To mitigate this, energy companies must explore energy-efficient AI algorithms, low-power computing hardware, and renewable-powered data centers to minimize AI's environmental impact.

Regulatory bodies, such as Public Utility Commissions (PUCs), ensure consumer protection and fair energy markets. Still, they must also keep pace with the rapid evolution of AI technologies. Overly rigid regulations could slow AI adoption, delaying time-to-value for innovations that could improve efficiency, lower costs, and enhance sustainability. Realizing AI's full potential requires collaboration between technology innovators, policymakers, and industry leaders. As the energy landscape continues to evolve, AI will remain a central force, empowering us to meet global energy needs responsibly and sustainably.

About the Author

Kent Landrum is a seasoned technology and business consultant with over two decades of experience helping organizations navigate complex challenges in the energy sector. He specializes in guiding companies through technology transformation, operational improvements, and data-driven decision-making, focusing on optimizing processes, integrating enterprise systems, and enhancing overall efficiency. His expertise spans energy trading, risk management, business intelligence, and supply chain operations, making him a trusted advisor to businesses seeking to modernize their operations. With a strong background in mergers and acquisitions, enterprise technology strategy, and process optimization, Kent has successfully led large-scale initiatives that drive measurable business value. His leadership style emphasizes collaboration, innovation, and practical solutions that align with organizational goals. Kent holds a B.S. in Computer Science and Economics from Trinity University and an M.A.A. in Organizational Development from the University of the Incarnate Word.

Email: kentblandrum@gmail.com

FROM FINANCIAL SERVICES TO GENERATIVE AI: A LEADER'S JOURNEY IN INNOVATION AND TRANSFORMATION

By Georg Langlotz
Head of GCRG AI Centre of Excellence
Zurich, Switzerland

AI Revolution in Finance: The $1 Trillion Transformation by 2030
—McKinsey, *The Executive AI Playbook*, 2024

The Foundation: Understanding AI in Financial Services

The Evolution of AI in Financial Services

Have you ever wondered how the world of finance transformed from dusty ledgers to cutting-edge algorithms? Picture this: a financial landscape where machines not only crunch numbers but predict market trends, detect fraud in milliseconds, and even chat with customers. Sound like science fiction? Welcome to the AI revolution in finance. Imagine stepping into a time machine and traveling back to the early 1980s. You'd find the first whispers of AI in finance— simple systems offering basic tax and financial advice. Fast-forward to today, and you're in a world where sophisticated machine learning algorithms power state-of-the-art generative AI solutions and AI agents. But how did we get here, and what does it mean for the future of finance?

At the heart of this transformation lies AI's superpower: the ability to process and analyze massive amounts of data at mind-boggling speeds. In a world where banks handle mountains of information daily, this capability is nothing short of revolutionary. It's like having a financial superhero that can spot hidden patterns, predict market movements, and make lightning-fast decisions with uncanny accuracy. And when it comes to critical operations like credit card fraud detection, AI's real-time analysis can be the difference between thwarting a fraudulent transaction and facing substantial losses.

But AI's impact on finance goes far beyond just number crunching. It's reshaping the very core of financial institutions, driving innovations that touch every aspect of the industry. From algorithmic trading that outpaces human reflexes to risk assessments that consider countless variables, AI is becoming the backbone of modern finance. And let's not forget about personalized services and compliance monitoring—areas where AI is making its mark in ways we could hardly imagine just a few years ago.

As we stand on the cusp of this AI-driven financial revolution, one can't help but wonder: what's next? Will AI democratize access to financial services, leveling the playing field for all? How will it shape the decision-making processes that drive the industry? One thing is certain—the future of finance is being written in lines of code, and AI is holding the pen.

The AI Ecosystem in Finance

As we dive deeper into this AI-powered financial wonderland, let's explore the four key realms where AI is making waves. It's like a high-tech ecosystem, each part working in harmony to create a financial world that's smarter, faster, and more intuitive than ever before.

First, we have personal productivity—imagine an AI assistant that understands your financial documents and manages your knowledge like a seasoned pro. It's like having a financial genius in your pocket, always ready to help you make sense of the complex world of finance.

In the domain of decision support, AI demonstrates its transformative potential. These advanced systems excel in market analysis, risk monitoring, and fraud detection, operating with precision and speed that surpasses human capabilities. AI's ability to process vast amounts of data and identify complex patterns enables financial institutions to anticipate market trends and potential risks with unprecedented accuracy. This enhanced predictive power provides a significant competitive advantage, allowing for more informed decision-making and strategic planning. The integration of AI in this area has effectively augmented human expertise, enabling analysts to focus on higher-level strategic tasks while AI handles complex data processing and pattern recognition. In business operations, AI catalyzes operational excellence. It significantly enhances efficiency by optimizing self-service support systems and streamlining organizational functions. This integration of AI technologies enables financial institutions to operate with increased agility, allowing them to adapt swiftly to market changes and emerging challenges. By automating routine tasks and providing data-driven insights, AI empowers institutions to allocate resources more effectively, reduce operational costs, and improve overall productivity. The result is a more responsive and resilient organizational structure capable of meeting the dynamic demands of the modern financial landscape.

In the customer experience domain, AI plays a pivotal role in transforming client interactions and service delivery. By leveraging advanced analytics and machine learning algorithms, AI enhances sales efficiency and improves product penetration rates. Moreover, it enables the creation of highly personalized financial offerings tailored to individual customer needs and preferences. This level of customization significantly elevates the customer experience, fostering stronger client relationships and increased loyalty. AI-driven systems can analyze vast amounts of customer data to provide insights and recommendations that rival those of experienced financial advisors, offering a scalable solution for personalized financial guidance. This technology-driven approach improves customer satisfaction and opens new revenue generation and market expansion avenues.

The financial services sector's investment in AI is experiencing significant growth, as evidenced by recent data from the World Economic Forum. Financial services firms invested $35 billion in AI technologies in 2023, with projections indicating this figure will reach $97 billion by 2027.[1] This substantial increase in investment underscores the industry's confidence in AI's transformative potential. The rationale behind this substantial investment is clear: approximately 70% of financial services executives believe AI will directly contribute to revenue growth in the coming years. McKinsey's analysis supports this optimism, which estimates that AI could generate up to $1 trillion of additional value annually for banks.[2] This projected value creation is expected to manifest through several key avenues:

- Enhanced service personalization
- Significant cost reductions
- New data-driven opportunities for financial growth and innovation

As the financial sector stands at the cusp of this AI-driven transformation, it raises crucial questions about the future of finance. We may be witnessing the emergence of a new era where human expertise and machine intelligence converge to create a more efficient, personalized, and profitable financial ecosystem.

Data: The Lifeblood of AI in Finance

Data has emerged as a critical asset in the evolving financial services landscape, fueling artificial intelligence engines and driving unprecedented innovation. The financial sector generates a staggering amount of data, with major institutions managing datasets on a petabyte scale. However, the actual value lies not merely in the quantity of data but in its quality and integration.

Consider this striking statistic: As of 2024, 77% of consumers utilize AI technologies for their banking and financial needs.[3] This widespread adoption has generated an invaluable repository of behavioral and transactional data. The challenge, however, lies in extracting meaningful insights from this vast information pool. This is where AI demonstrates its transformative potential. By leveraging advanced algorithms and machine learning techniques, AI systems can process and analyze this enormous volume of data, converting raw information into actionable insights. These insights, in turn, inform strategic decision-making processes across financial institutions.

The implications of this data-driven approach are far-reaching. Financial services providers can now offer more personalized products and services, enhance risk assessment models, and improve operational efficiency. Moreover, AI-powered data analysis enables institutions to identify market trends, predict customer behavior, and develop innovative financial solutions with unprecedented accuracy. As we navigate this data-rich environment, several critical questions emerge:

- How can financial institutions optimize their data infrastructure to leverage AI capabilities fully?
- What measures are necessary to ensure data privacy and security in this increasingly connected financial ecosystem?
- How will the growing reliance on AI-driven insights reshape the competitive landscape in financial services?

The answers to these questions will likely shape the future trajectory of the financial industry. As AI evolves and data volumes grow exponentially, financial institutions that successfully harness these technologies will be well positioned to lead in an increasingly digital and data-centric world.

Crafting a Comprehensive AI Strategy and Roadmap

A well-crafted AI strategy and roadmap are essential for organizations looking to harness the power of artificial intelligence effectively. This strategic approach ensures that AI initiatives are aligned with corporate objectives and deliver quantifiable financial outcomes.

Aligning AI Initiatives with Business Goals

The foundation of a successful AI strategy lies in its alignment with overall business objectives. This requires:

- Clearly defining strategic business outcomes that artificial intelligence can influence, such as enhancing customer experiences or improving operational efficiencies
- Focusing AI investments on areas with the highest potential for return, considering both short-term impacts and long-term transformation
- Integrating AI initiatives within the fabric of overall business strategies to complement and accelerate broader business goals

Prioritizing Use Cases

A critical step in developing an AI roadmap is the prioritization of use cases:

- Evaluate potential AI applications across different business units and functions.
- Assess each use case based on its potential impact, feasibility, and alignment with strategic objectives.
- Consider factors such as data availability, technical complexity, and potential ROI when prioritizing initiatives.

Defining Success Metrics

To measure the effectiveness of AI initiatives, it's crucial to establish clear success metrics:

- Develop specific, measurable, and time-bound objectives for each AI project.
- Align these metrics with broader business KPIs to demonstrate tangible value.
- Regularly track and report on these metrics to stakeholders, ensuring accountability and enabling data-driven decision-making.

Securing Executive Sponsorship

Executive buy-in is critical for driving organizational change and ensuring the success of AI initiatives:

- Present the AI strategy to stakeholders, clearly communicating the benefits, costs, and expected results.
- Secure the necessary budget and resources to implement the strategy effectively.
- Engage leadership in championing AI initiatives and demonstrating commitment through clear communication and resource allocation.

Outlining Progressive Stages of AI Maturity

A comprehensive AI roadmap should outline the journey from initial pilot programs to enterprise-wide deployment:

1. Exploring: identify potential AI use cases and conduct initial experiments.

2. Planning: develop a formal AI strategy and plan for broader deployments.

3. Implementing: transition from proofs of concept to full-scale AI deployments.

4. Scaling: expand successful AI initiatives across the organization.

5. Realizing: fully integrate AI into business processes and realize its transformative potential.

Organizations can craft a dynamic AI strategy and roadmap that propels innovation, enhances efficiency, and delivers concrete business value by adopting a structured yet flexible approach. This strategic framework ensures that AI initiatives are not isolated technological projects but rather integral components of the organization's broader digital transformation journey.

Crucially, developing this strategy should be iterative, incorporating continuous feedback and insights from key stakeholders across the organization. This agile methodology allows for rapid adjustments and refinements, ensuring the AI strategy remains aligned with evolving business needs and technological advancements. By fostering ongoing dialogue and collaboration between technical teams, business units, and leadership, organizations can create a more responsive and effective AI implementation plan that adapts to changing circumstances while staying true to core business objectives.

Building a Robust Data Infrastructure

A robust data infrastructure is the backbone of successful AI implementation in any organization. It combines cutting-edge cloud architecture with sophisticated data management practices to build a foundation for advanced analytics and AI model deployment. Here are the key components and considerations:

Cloud-Based Architecture

- Leverage cloud platforms (e.g., AWS, Azure, Google Cloud) for scalability and flexibility.
- Implement a hybrid or multi-cloud strategy to optimize performance and cost.
- Utilize containerization and microservices for modular, scalable applications.

Data Management and Governance

- Establish comprehensive data quality frameworks to ensure accuracy and consistency
- Implement master data management (MDM) practices for a single source of truth.
- Develop clear data governance policies, including data ownership and access controls.

Data Lakes and Warehouses

- Create centralized data repositories to store structured and unstructured data.
- Implement data cataloging and metadata management for easy discovery and use.
- Ensure data lineage tracking for regulatory compliance and auditing purposes.

API Development and Integration

- Design and implement RESTful APIs for seamless data access and integration.
- Use API gateways to manage, secure, and monitor API traffic.

- Adopt microservices architecture for modular and scalable data services.

Real-Time Processing Capabilities

- Implement stream processing technologies.
- Utilize in-memory databases for high-speed data processing.
- Develop event-driven architectures for real-time data updates and actions.

Data Security and Privacy

- Implement robust encryption for data at rest and in transit.
- Use identity and access management (IAM) tools for fine-grained access control.
- Ensure compliance with data protection regulations (e.g., GDPR, CCPA).

Scalable AI Model Deployment

- Utilize containerization (e.g., Docker) and orchestration (e.g., Kubernetes) for model deployment.
- Implement MLOps practices for continuous integration and delivery of AI models.
- Develop model monitoring and retraining pipelines for maintaining model performance.

Data Visualization and Reporting

- Implement business intelligence (BI) tools for data exploration and visualization.
- Develop self-service analytics capabilities for business users.
- Create dashboards and reports for real-time monitoring of key metrics.

By focusing on these areas, organizations can build a robust data infrastructure that supports current AI initiatives and provides the flexibility and scalability needed for future growth and innovation.

Cultivating AI-Ready Talent and Culture

Organizations must foster an AI-ready culture through systematic talent development and transformation. This begins with leadership championing AI initiatives and demonstrating commitment through resource allocation and clear communication. The transformation requires three key elements:

1. Develop comprehensive reskilling programs that blend technical AI knowledge with domain expertise, enabling employees to identify and implement AI opportunities within their functions.

2. Establish centers of excellence that combine recruited AI specialists with domain experts, creating hubs for knowledge sharing and innovation acceleration.

3. Implement structured learning paths and reward systems to encourage experimentation and cross-functional collaboration.

Success in building an AI-ready organization depends on creating feedback loops between business units, maintaining continuous learning programs, and establishing clear career progression paths for AI talent. Regular AI literacy workshops, hackathons, and innovation challenges help embed AI thinking into daily operations while fostering a culture of experimentation and data-driven decision-making.

Implementing an AI-Integrated Operating Model

Integrating AI capabilities into business processes requires a purpose-built operating model that balances innovation with operational stability. This model should be designed to leverage AI's potential while aligning with the organization's strategic goals and existing workflows.

Agile Delivery Frameworks

- Agile methodologies: implement scrum, kanban, or hybrid approaches to enable rapid iteration and continuous improvement of AI solutions.
- Cross-functional teams: form teams that blend AI expertise with domain knowledge to ensure AI solutions address real business needs.
- Sprint-based development: use short development cycles to prototype, test, and refine AI models and applications quickly.

Clear Accountability Structures

- Defined roles and responsibilities: clearly outline the roles of data scientists, engineers, business analysts, and domain experts in AI projects.
- Clear governance framework: create a governance structure to oversee AI initiatives and ensure alignment with ethical guidelines and regulatory requirements.
- Key performance indicators (KPIs): develop and track AI-specific KPIs to measure the impact and effectiveness of AI implementations.

Seamless Coordination

- Collaborative platforms: utilize tools that facilitate communication and collaboration between technical and business teams.
- Regular touchpoints: schedule frequent meetings between AI teams and business units to ensure ongoing alignment and address challenges promptly.

- Knowledge sharing sessions: organize workshops and seminars to bridge the knowledge gap between technical and non-technical staff.

Rapid Experimentation

- Innovation labs: establish dedicated spaces for experimenting with new AI technologies and use cases.
- Sandbox environments: create isolated testing environments to safely experiment with AI models without risking operational systems.
- Fail-fast culture: encourage teams to test hypotheses and learn from failures to accelerate innovation quickly.

Maintaining Operational Stability

- Robust testing protocols: implement comprehensive testing procedures to ensure reliable and secure AI solutions before deployment.
- Gradual rollout strategy: use phased implementations to minimize disruption and allow for real-world validation of AI solutions.
- Continuous monitoring: employ AI-powered monitoring tools to detect and address potential issues in real-time.

Scalability and Flexibility

- Cloud-based infrastructure: leverage cloud platforms to ensure scalability and flexibility in AI deployments.
- Microservices architecture: design AI systems using microservices to enable easier updates and scaling of individual components.
- API-first approach: develop APIs for AI services to facilitate integration with existing systems and future applications.

By implementing this AI-integrated operating model, organizations can effectively embed AI capabilities into their core business processes, fostering innovation while maintaining the stability required for day-to-day operations. This approach enables companies to remain agile despite rapid technological advancements while delivering tangible business value through AI initiatives.

Governance and Ethics

A comprehensive AI governance framework in the financial services industry (FSI) demands a unique approach, addressing challenges that set it apart from other sectors. At its core, this framework orchestrates responsible development and deployment through multiple interconnected layers, emphasizing transparency and fairness due to the significant impact these models have on individuals' lives and the stringent regulatory requirements in the industry.

The foundation of this framework is built upon robust model risk management, which encompasses rigorous performance monitoring, advanced bias detection mechanisms, and regular model validation protocols. These elements are crucial in maintaining the integrity and reliability of AI systems in financial services, where decisions can have far-reaching consequences for customers and stakeholders. Supporting this foundation is a strong project delivery governance structure that ensures systematic development cycles, stringent quality controls, and clear accountability structures, all essential in an industry where regulatory compliance is paramount.

Stakeholder governance is integrated through established committees and decision-making processes, carefully balancing the drive for innovation with the imperative of risk management. This layer is critical in FSI, where the need for transparency often limits the techniques employed in AI development. The framework incorporates automated fairness assessments and ethics review boards for sensitive AI applications, addressing the crucial need for equitable treatment in financial decision-making processes.

To meet the industry's unique challenges, the governance structure includes clear escalation paths for model degradation, standardized validation procedures with comprehensive documentation, and regular independent audits and compliance reviews. These measures ensure that AI systems remain reliable and ethical and adhere to the financial sector's strict regulatory standards. Transparent reporting mechanisms for stakeholders further reinforce trust and accountability, which are essential in an industry where public confidence is paramount.

This multi-layered, FSI-specific approach to AI governance ensures that artificial intelligence systems in finance remain cutting-edge, efficient, reliable, ethical, and compliant. By addressing the unique challenges of transparency and fairness head-on, this framework maintains stakeholder trust and regulatory adherence while enabling financial institutions to harness the full potential of AI technologies responsibly.

Leadership in the AI Era: Balancing Technology and Humanity

The AI revolution in finance has ushered in a new era of leadership that demands a delicate balance between technological prowess and human touch. Today's financial leaders find themselves at the intersection of innovation and empathy, tasked with navigating a landscape where algorithms and emotions carry equal weight. Imagine a tightrope walker balancing technical acumen in one hand and emotional intelligence in the other. This modern financial leader adeptly maneuvers through an AI-driven environment while maintaining ethical governance. As employees increasingly seek connection in a digital world, leaders face the challenge of nurturing relationships while driving technological transformation.

But how do we address the elephant in the room—employee concerns about AI integration? Successful leaders are tackling this head-on, fostering a culture of trust and psychological safety. They're not just implementing AI; they're weaving a narrative where AI augments human potential rather than replaces it. Transparency becomes their watchword, as they guide their organizations through digital transformation, focusing on human development.

In this new landscape, ethics takes center stage. It's no longer about mere compliance but cultivating a culture of responsibility and trust. Leaders are now architects of robust governance frameworks, ensuring AI decision-making is transparent, fair, and secure. They're tackling algorithmic bias head-on, recognizing that AI ethics isn't just a moral imperative—it's a competitive differentiator.

While AI enhances leadership capabilities through real-time feedback and workflow optimization, the essence of leadership remains deeply human. It's about fulfilling those fundamental psychological needs—autonomy, competence, and relatedness. The challenge? Striking the right chord between leveraging AI's potential and preserving the irreplaceable human elements that fuel creativity and innovation.

We see a dynamic and multifaceted leadership paradigm as we peer into the future. Tomorrow's leaders will be tech-savvy visionaries, driving organizational change and fostering cross-functional collaboration. They'll be nimble adapters, encouraging their teams to embrace new technologies with open minds. But more than that, they'll be the guardians of ethical AI use, championing robust data protection and governance frameworks.

The AI leader of tomorrow is not just a manager of technology, but a curator of human potential. They'll seamlessly blend technological expertise with human-centric leadership, propelling their organizations into a future where AI and human capabilities are symbiotically maximized.

In this brave new world of finance, the leaders who thrive will be those who can harness the power of AI while never losing sight of the human heart that beats at the center of every successful organization. They'll be the ones who turn the challenges of the AI era into opportunities for growth, innovation, and meaningful connections.

Generative AI: The Next Frontier in Financial Services

Understanding Generative AI

Generative AI represents a transformative force in financial services, introducing a revolutionary paradigm that transcends traditional technological advancements. Its unique ability to simultaneously drive productivity improvements while enabling creative problem-solving marks an unprecedented breakthrough in enterprise technology. Technology's nondeterministic architecture navigates ambiguity and complexity in ways previously reserved for human cognition, fundamentally transforming how financial institutions approach both routine operations and complex challenges.

This innovation emerges at a critical juncture as financial institutions face mounting pressure to accelerate digital transformation. Driven by evolving customer expectations, competitive disruption from FinTech startups, and rapid technological advancement, traditional banks must adapt to survive. Modern consumers demand seamless, personalized banking experiences while neobanks continue to redefine industry standards with innovative digital-first solutions.

Implementing Generative AI: Challenges and Opportunities

Implementing generative AI in the financial services sector presents a complex landscape of opportunities and challenges, with data privacy and security at the forefront of concerns. Given the sensitive nature of information held by financial institutions, safeguarding customer data is paramount, as trust forms the bedrock of success in this industry. While generative AI offers significant potential for enhancing operational efficiency, personalizing customer experiences, and advancing risk management capabilities, it also introduces substantial hurdles that organizations must carefully navigate.

Chief among these challenges is the imperative to maintain stringent data privacy protocols and robust security measures, ensuring that AI does not compromise the confidentiality of client information. Moreover, the inherent opacity of many AI techniques poses significant obstacles regarding transparency and regulatory compliance. Financial institutions must grapple with the difficulty of debugging these complex systems, monitoring for bias, and identifying root causes of unexpected behaviors—all crucial aspects for meeting regulatory requirements and maintaining public trust. Additional challenges include effective model governance, acquisition of specialized AI talent, and seamless integration with legacy systems. To successfully harness the power of generative AI, financial organizations must address these multifaceted challenges while adhering to regulatory standards and ensuring ethical AI deployment. This delicate balancing act requires a strategic approach prioritizing data protection, transparency, and responsible innovation to fully capitalize on AI's transformative potential in finance.

The convergence of generative AI with other technologies presents crucial opportunities for modernization in an increasingly competitive and technology-driven financial landscape, where innovation is no longer optional but a strategic imperative. Financial institutions that successfully implement generative AI while addressing its challenges will gain significant competitive advantages in operational efficiency, customer service, and market responsiveness.

Future Trends: Preparing for the Next Wave of AI in Finance

As we look ahead to 2025 and beyond, the financial services industry stands on the cusp of a transformative era driven by artificial intelligence. The convergence of AI and finance is not just reshaping existing paradigms but creating new possibilities that will redefine how we approach banking, investment, and risk management.

The advent of agentic AI marks a paradigm shift in artificial intelligence, ushering in a new era of autonomous systems capable of independent decision-making, execution, and adaptation. This revolutionary technology, based on large language models, is poised to redefine the landscape of financial services, offering unprecedented levels of automation, analytical prowess, and personalization.

Enhanced Automation

Agentic AI systems can autonomously handle complex, multi-step processes without constant human oversight. In financial operations, this translates to:

- Streamlined back-office operations
- Automated compliance checks and regulatory reporting
- Dynamic risk assessment and management

Advanced Predictive Analytics

By continuously learning from outcomes and adapting to changing conditions, agentic AI elevates predictive capabilities:

- Real-time market analysis and trend forecasting
- Proactive fraud detection and prevention
- Anticipatory customer behavior modeling

Hyper-Personalized Services

Agentic AI's ability to process vast amounts of data and make nuanced decisions enables:

- Tailored financial advice and product recommendations
- Adaptive investment strategies aligned with individual goals
- Personalized customer interactions across all touchpoints

While the potential of agentic AI is immense, its implementation raises several critical considerations:

- Control and oversight: as AI systems become more autonomous, establishing appropriate human oversight mechanisms becomes crucial to ensure alignment with organizational goals and values.
- Accountability: determining responsibility for decisions made by agentic AI systems presents complex legal and

ethical challenges, particularly in high-stakes financial operations.

- Ethical implementation: financial institutions must navigate ethical considerations, including:

  o Algorithmic fairness and bias mitigation in AI, particularly in the context of generative AI, have become critical concerns for financial institutions. The challenge has evolved beyond just addressing biases in decision-making data; it now encompasses the inherent biases within the foundational models themselves. These biases, originating from the vast datasets used in pre-training, can insidiously influence subsequent decisions, even after fine-tuning by financial institutions. To combat this, organizations are implementing multi-faceted approaches. These include diversifying training data, designing bias-aware algorithms, and employing techniques like adversarial training and fairness constraints. Continuous monitoring and evaluation are essential, as is fostering diversity in development teams to bring varied perspectives to bias identification. Transparency in AI systems, through explainable AI techniques, aids in uncovering hidden biases. Financial institutions also adopt strategies such as de-biasing data, creating algorithmic audits, and establishing internal committees to review AI decision-making systematically. While these efforts are crucial, it's important to note that mitigating bias often involves a trade-off with model performance, necessitating a balanced approach. As GenAI continues to evolve, vigilance and adaptability in bias mitigation strategies remain paramount to ensure fair and ethical AI applications in finance.

  o Transparency in AI decision-making processes

  o Data privacy and security

- Regulatory compliance: the deployment of agentic AI must adhere to evolving regulatory frameworks, necessitating close collaboration between financial institutions and regulatory bodies.

As agentic AI continues to evolve, it promises to revolutionize financial services by enabling more sophisticated, efficient, and personalized operations. However, realizing its full potential will require careful navigation of ethical, regulatory, and operational challenges to ensure responsible and beneficial implementation. We expect to see accelerated adoption of generative AI and agentic AI across the sector, streamlining processes such as loan approvals, fraud detection, and customer support. AI-driven tools are processing transactions up to 90% faster than traditional methods, and by 2030, AI is projected to save banks billions annually through enhanced productivity and operational efficiencies.

The future of finance will also see a shift towards hyper-personalization, with AI and advanced analytics enabling institutions to deliver tailored financial products and services at scale. From bespoke savings plans based on individual spending habits to AI-powered investment advice aligned with personal goals, the customer experience in finance is set to become more personalized than ever before.

The Changing Role of Financial Professionals

Integrating AI into finance has redefined the skills and responsibilities of financial professionals. While more than 50% of financial sector jobs have a high potential for automation, this presents an opportunity for evolution and growth rather than doom for finance professionals.

Financial professionals are now expected to be tech-savvy and data- and AI-literate, understanding how AI algorithms work, interpreting complex data patterns, and using AI-driven insights to make strategic decisions. This shift has given rise to hybrid roles that combine financial expertise with technological fluency.

The future of finance lies in the symbiosis between human expertise and AI capabilities. Financial professionals who can effectively collaborate with AI systems, leveraging their strengths while compensating for their limitations, will be the most valuable assets in the industry.

Conclusion: Embracing the AI-Driven Future of Finance

As we stand at the precipice of a new era in financial services, artificial intelligence emerges not as a fleeting trend but as a catalyst for profound transformation. The evolution from traditional financial practices to AI-driven innovation has been a journey of both excitement and complexity, demanding a nuanced approach that balances vision, strategy, and execution.

Integrating AI across the financial spectrum—from fraud detection and risk management to personalized banking and investment strategies—has demonstrated its potential to revolutionize efficiency, accuracy, and customer experience. Forward-thinking financial institutions embracing this technology are reaping substantial benefits: streamlined operations, enhanced decision-making processes, and the ability to offer bespoke services tailored to individual client needs.

However, this path of innovation is not without its hurdles. As we've explored, concerns surrounding data privacy, ethical considerations, and the necessity for robust governance frameworks remain at the forefront of the AI revolution in finance. The industry must navigate these challenges with prudence, ensuring that the pursuit of innovation does not compromise trust, security, or regulatory compliance.

The advent of agentic AI and advanced generative models represents the next frontier in financial services. These cutting-edge technologies promise to automate complex processes further, provide deeper insights, and enable hyper-personalized financial solutions. Yet, they also underscore the critical importance of maintaining human oversight and ethical considerations in AI deployment.

For financial leaders charting a course into this AI-driven future, fostering an AI-ready culture within their organizations is paramount. This entails investing in technology and people—developing talent, promoting cross-functional collaboration, and cultivating a mindset that embraces continuous learning and adaptation. The financial institutions poised to thrive in this AI-driven landscape will be those that strike an optimal balance between technological advancement and human expertise. They will leverage AI not as a replacement for human capabilities, but as an augmentation, creating synergistic relationships that drive innovation and deliver unparalleled value to customers.

In essence, the AI revolution in finance transcends mere technological adoption; it represents a fundamental reimagining of financial services. It offers an unprecedented opportunity to create more inclusive, efficient, and responsive financial systems that can better serve the evolving needs of individuals and businesses alike. As we embrace this AI-driven future, the financial industry stands on the brink of unlocking extraordinary opportunities for growth, innovation, and societal impact. The journey ahead is undoubtedly complex and multifaceted, but the potential rewards are transformative for those willing to embrace change and lead with vision. The future of finance is unfolding before us, powered by the limitless possibilities of artificial intelligence.

Chapter Endnotes

1.https://reports.weforum.org/docs/WEF_Artificial_Intelligence_in_Financial_Services_2025.pdf

2. https://www.mckinsey.com/capabilities/quantumblack/our-insights/the-executives-ai-playbook?page=industries/banking/

3. https://www.allaboutai.com/resources/ai-statistics/finance/

About the Author

Georg Langlotz is a distinguished leader in data analytics and digital transformation expert with over two decades of experience driving innovation and operational excellence across the financial services industry. Currently serving as the head of GCRG AI Centre of Excellence at UBS, Georg has established himself as a visionary who leverages cutting-edge technologies to revolutionize business processes and customer experiences.

Throughout his career, Georg has held pivotal roles at major financial institutions, including data product and delivery manager at Credit Suisse, director of operations Switzerland at Colt Group plc, and head of global platform engineering at Man Group. His expertise spans many areas, including front office digital business transformation, operational risk solutions, customer experience management, data analytics and AI implementation, machine learning governance, and robotic process automation. Known for his analytical, structured, and entrepreneurial mindset, Georg consistently excels in leading global teams and transforming departments to exceed customer expectations. His approach combines strategic thinking with hands-on experience in process excellence and data-driven decision-making, making him a valuable asset in the rapidly evolving landscape of financial technology.

A committed lifelong learner, Georg has enhanced his skills through various educational pursuits, including a digital leader masterclass at the University of Edinburgh, an MBA focusing on corporate strategy from the University of Strathclyde, and studies in digitalization at Boston University and economics at HWV ZH. Georg's passion for leveraging disruptive technologies to drive business model innovation, his extensive experience, and strategic vision position him as a thought leader in digital transformation within the financial services industry. His ability to navigate complex challenges and drive meaningful change continues to shape the future of data analytics and AI applications in finance.

Email: glanglotz@gmail.com

LinkedIn: https://www.linkedin.com/in/georg-langlotz/

CHAPTER 16

THE AI REVOLUTION IN HEALTHCARE

By Glenn Loomis, MD, MS, FAAFP
Founder, Query Health; Healthcare Leader, AI
Henderson, Nevada

Artificial intelligence will not replace doctors. But doctors who use artificial intelligence will replace those who don't.

—Dr. Bertalan Meskó, The Medical Futurist

Introduction: The Transformation Is Here

As someone who has watched the healthcare industry evolve over the years, I can confidently say that we are in the midst of a revolution—one unlike any in its history. Artificial intelligence (AI) is no longer just a buzzword or a futuristic dream. It's here, reshaping how we diagnose, treat, and care for patients. What makes this transformation so remarkable is AI's ability to learn, adapt, and perform tasks with precision that rivals or exceeds human ability.

This change hasn't happened overnight. Instead, it's been a gradual integration of AI into healthcare systems, addressing inefficiencies and solving problems that once seemed insurmountable. From radiology to patient monitoring and even mental health, AI is making an impact that's impossible to ignore. Despite this amazing progress, I believe we're only scratching the surface of what's possible. The future promises even greater advancements, and the question we face isn't whether AI will transform healthcare—it's how we can best prepare for and adapt to it.

In this chapter, I'll explore some of the incredible ways AI is already making a difference, dive into the exciting applications on the horizon, and share thoughts on how healthcare professionals, systems, and patients can adapt to thrive in this rapidly evolving landscape. Let's start by looking at what's happening right now—because the revolution isn't coming. It's already here!

Current Applications of AI in Healthcare

AI is already transforming the healthcare industry in ways that felt like science fiction just a few years ago. Let me share some examples of how AI is changing the game right now.

Administrative Tasks

If you've ever felt frustrated by the inefficiencies of healthcare administration, you're not alone. I've seen firsthand how administrative burdens—scheduling, documentation, insurance claims—can drain valuable time and energy. That's where AI comes in. Tools like Nuance's Dragon Ambient eXperience (DAX), which uses natural language processing (NLP), automate the transcription of physician notes. Imagine a world where doctors focus more on patients and less on paperwork. That's the reality AI is creating.

AI is also transforming scheduling. For example, I recently spoke with a clinic administrator who shared how machine learning algorithms used by Zocdoc predict patient no-show rates, allowing them to optimize appointments. And let's not forget prior authorization—a process many providers dread. AI technology from companies like Thoughtful now automates much of it, cutting down approval times and ensuring patients get the care they need faster.

Radiology

When I think about the impact of AI in radiology, I'm reminded of a story about a radiologist who was initially skeptical about AI. That changed when an AI system flagged a tiny tumor in a mammogram that even the radiologist admitted they might have missed. Tools like these, powered by millions of medical images, are revolutionizing how we detect anomalies. Google Health's AI system, for instance, has demonstrated extraordinary precision in identifying breast cancer. Zebra Medical Vision's AI algorithms analyze imaging data to detect conditions like osteoporosis, liver disease, and coronary artery disease. Aidoc's AI solutions analyze CT scans in real-time to detect life-threatening conditions like intracranial hemorrhages. These systems don't replace radiologists but act as an indispensable second set of eyes, reducing oversight errors and enabling earlier, life-saving interventions.

Predictive Analytics

One of the most exciting aspects of AI is its ability to predict health issues before they happen. Imagine being able to intervene before a patient develops sepsis or prevent a hospital readmission by identifying at-risk patients in advance. Jvion's predictive AI identifies patients at risk for complications like sepsis or readmission. At a hospital in Texas, Jvion's platform reduced readmission rates by 20% in high-risk patients, demonstrating the potential of AI to save both lives and costs. Hospitals using Epic's predictive analytics can flag patients at high risk of sepsis hours before symptoms become apparent. One hospital reported a 30% reduction in sepsis-related mortality after implementing these tools. Other examples include Clarify Health, whose AI models use patient and population data to predict outcomes and optimize care pathways, reducing overall treatment costs. I've seen predictive models in action, and the results are often remarkable. Hospitals using these tools have not only improved patient outcomes but also saved millions in costs.

Virtual Health Assistants

AI-powered virtual health assistants are another area of incredible innovation. I remember testing an app like CareAI and being amazed at its ability to triage symptoms, recommend care pathways, and even answer medical questions. Another example is HealthTap that uses AI-powered assistants that provide symptom checkers, personalized health advice, and access to a network of doctors for follow-ups. A patient I spoke with used HealthTap during a late-night emergency to determine whether they needed urgent care, saving them an unnecessary trip to the ER. These 24/7 assistants provide support that empowers patients and reduces the burden on healthcare providers by addressing non-urgent queries and routine care needs.

Billing and Coding

Anyone who's worked in healthcare knows how tedious and error-prone billing and coding can be. AI systems, such as those developed by Optum360, are automating these processes, ensuring accuracy, and speeding up reimbursement cycles. A health system implemented nThrive's AI platform to automate the coding of medical procedures and saw a 40% reduction in claim denials and improved cash flow, highlighting AI's potential to streamline the revenue cycle. I've also spoken with medical practices that have cut down their claim denials significantly thanks to AI, freeing staff to focus on more critical tasks.

Remote Patient Monitoring

The rise of wearable devices and internet of things (IoT) technology has brought AI into the realm of remote patient monitoring. Current Health offers an AI-powered wearable that monitors vital signs like heart rate and oxygen levels in real time. This has been shown to reduce hospital admissions for high-risk patients by 25% through catching potential issues early. Biofourmis uses AI-driven wearable devices to monitor patients recovering from heart surgery. The system flags early warning signs of complications, enabling timely interventions and leading to fewer readmissions and better recovery outcomes. I've seen how these tools are changing lives, particularly for patients with chronic conditions. Even something as simple as a smartwatch can make a difference. I once met a patient whose Apple Watch alerted them to an irregular heartbeat, leading to an early diagnosis of atrial fibrillation. Stories like this highlight how AI in patient monitoring is enabling proactive, personalized care.

Mental Health

AI's role in mental healthcare is extremely important. Studies show patients will tell the computer things they will not tell a provider physician in person. I've seen how virtual platforms like Woebot use NLP to provide real-time cognitive behavioral therapy (CBT) techniques. Users have praised it for providing nonjudgmental, round-the-clock support. In one study, 70% of Woebot users reported reduced symptoms of anxiety within two weeks. Another tool, EllieGrid uses AI to manage medications for patients with mental health conditions, ensuring adherence to prescribed regimens. Patients report feeling more in control of their health, and providers have seen improved outcomes as a result. These tools are helping people manage anxiety, depression, and stress in ways that feel supportive and accessible. Beyond direct patient care, new AI tools are also detecting mental health issues through speech patterns and social media behavior, offering hope for earlier interventions.

Coming Applications of AI in Healthcare

Looking ahead, the possibilities for AI in healthcare are nothing short of extraordinary. Let me take you on a journey into what's coming.

Physician Services

Imagine a world where AI handles routine processes or cases, leaving physicians free to focus on complex cases. Here are some examples: Buoy Health is an AI-powered platform which provides virtual consultations by analyzing symptoms and recommending care. Infermedica's AI platform provides clinical decision support, guiding physicians with diagnostic suggestions based on patient symptoms and history. I've seen prototypes of virtual consultation systems that put all of this together, allowing AI to analyze symptoms, medical history, and lab results to recommend treatment plans. This could be a game-changer for reducing the impact of the physicians' shortage and improving access to care in underserved areas.

Advanced Robotic Surgery

Both the Hugo Robotic-Assisted Surgery (RAS) and DaVinci RAS platforms are combining AI and robotics to provide enhanced surgical capabilities. I had the chance to observe a robotic surgery demonstration powered by AI. What struck me was the robot's ability to adapt in real-time, using machine learning to anticipate complications. This kind of technology has the potential to make surgeries safer and more precise, reducing recovery times and improving outcomes.

Autonomous Diagnostics

One of the most exciting developments is the idea of autonomous diagnostics. Imagine an AI system that can read imaging data, lab results, or genetic information with little to no human intervention. Eyenuk's AI solution detects diabetic retinopathy through retinal imaging, providing diagnostic results without requiring a specialist. This system is already being deployed in rural areas, where access to ophthalmologists is limited. I've seen early versions of these types of systems in action, and the potential to deploy these tools in rural or resource-poor areas and democratize access to diagnostic care is staggering.

Tailored Precision Medicine

The future of medicine is personal. I've heard researchers describe their vision of AI analyzing genetic, lifestyle, and environmental data to create treatment plans tailored to individual patients. Deep Genomics uses AI to analyze genetic data and predict which drug therapies will work best for individual patients. Tempus uses AI to analyze genomic data and recommend personalized cancer treatments. By identifying specific genetic mutations, Tempus has helped oncologists develop more effective therapies for patients. The concept of having AI search the data for "patients like me" and return a therapeutic plan tailored to an individual's exact medical history and current circumstances is the holy grail. This isn't just about better outcomes—it's about a future where medicine works for each person's unique circumstances—optimizing patient outcomes while minimizing side effects.

New Diagnostics

AI is breaking new ground in diagnostics, from liquid biopsies to tools that can analyze subtle changes in speech. Grail's liquid biopsy tests use AI to detect early-stage cancers by analyzing biomarkers in blood samples. Early trials show promise in identifying cancers that traditional methods often miss. AI is also advancing Alzheimer's diagnosis, with researchers at MIT using AI to detect subtle changes in speech patterns and researchers at King's College London using AI to detect changes in brain scans, both predicting Alzheimer's cognitive decline years before symptoms appear. I've seen the promise of these technologies, and it's hard not to feel optimistic about the impact they will have.

Preparing to Thrive in AI Healthcare

To embrace this revolution, we must prepare. Healthcare professionals, patients, and policymakers all have roles to play, and several steps must be taken:

Embrace Lifelong Learning

For me, the journey of understanding AI has been eye-opening. As healthcare professionals, we need to embrace lifelong learning, equipping ourselves with the skills to use AI effectively. Programs like Stanford's AI for Healthcare Bootcamp equip clinicians with the knowledge to integrate AI into their practice, ensuring they remain at the forefront of innovation. Medical schools must also lead the way, by integrating AI into their curriculums.

Build Ethical Frameworks

"With great power comes great responsibility." I've had many conversations about the ethical challenges of AI, from data privacy to algorithmic bias. Addressing these issues is essential to ensure AI benefits everyone equitably. Partnership on AI, a group of healthcare stakeholders, develops guidelines to ensure fairness, transparency, and accountability in AI applications. IBM has established an internal ethics board to review the fairness, transparency, and accountability of its AI systems. Similar frameworks are needed across the industry to address concerns about bias and privacy.

Foster Collaboration

Collaboration is key. I've seen firsthand how partnerships between technologists, healthcare providers, and policymakers can drive meaningful innovation. Amazon, Google, and Microsoft all partner with healthcare organizations to develop scalable AI tools for disease prevention and treatment, showcasing the value of cross-sector collaboration. The Coalition for Healthcare AI (CHAI) is working with all types of stakeholders to develop standards for developing and using healthcare AI. Together, we can ensure AI tools are both effective and ethical.

Invest in Infrastructure

To harness AI's potential, we need robust infrastructure. Organizations at the forefront of healthcare AI, like Cleveland Clinic and Mayo Health System, have made huge investments in secure data storage, high-speed connectivity, and standardized formats for seamless data sharing to improve patient care. Unfortunately, healthcare has lagged other sectors in digitizing our processes, and this will impact our ability to capitalize on the AI revolution. Leaping this chasm is crucial for future success.

Focus on Patient Engagement

Finally, patients must be part of this journey. Mayo Clinic's Patient Portal educates patients on how AI enhances their care, fostering trust and engagement with AI-powered tools. Ada educates users about AI's capabilities while providing symptom-checking tools, helping to build trust and engagement with digital health solutions. I believe this type of education for patients about AI's benefits can alleviate fears and build trust, paving the way for wider adoption.

Conclusion

The AI revolution in healthcare is no longer a distant vision—it's happening right now. As I reflect on the AI revolution in healthcare, I'm struck by its transformative potential. From the tools we're already using to the groundbreaking advancements on the horizon, AI is not just improving healthcare—it's redefining it. And as emerging applications like autonomous diagnostics and tailored precision medicine come to fruition, the potential to save lives and improve outcomes is boundless.

To thrive in this new era, we must prepare by embracing lifelong learning, addressing ethical challenges, fostering collaboration, and investing in infrastructure. Above all, we must remember that AI is not here to replace the human touch in healthcare—it's here to amplify it.

Note: All company names are illustrative, and the author has not received any type of compensation for their inclusion.

About the Author

Glenn A. Loomis, MD, MS, FAAFP, is an American physician, executive, and entrepreneur. He has a 30-plus-year career of teaching medicine, leading physicians, and developing healthcare technology and virtual care. His organizations are known for operational excellence and have won prestigious awards, including the Malcolm Baldrige award for quality, the AMGA Acclaim Award, Health Finance and Management Association Revenue Cycle Award, IBM Watson Health Advantage Award for Population Health, and the Most Wired Health System and Physician Group Award.

Glenn received his BS and MD degrees from The Ohio State University and is an avid supporter of the university. He completed his MS in Healthcare Management at the University of Texas at Dallas and fellowships in medical education and healthcare policy. He served as a leader in the American Medical Association (AMA), American Medical Group Association (AMGA), and the American Academy of Family Physicians (AAFP), including authoring the AMA's initial white paper on AI in healthcare. He was named a fellow of the AAFP and of the New York Academy of Medicine. He has received multiple awards for teaching, leadership, and public service. In 2014, he was named the Citizen Doctor of the Year by the Kentucky Academy of Family Physicians. He also served four years on active duty as a physician in the US Air Force.

In 2024, Dr. Loomis founded Query Health, Inc with a vision to provide AI-enhanced medicine to the planet. This company is creating an AI physician agent to automates many daily physician tasks, such as speaking with patients, gathering their medical information, and presenting that information to their physician with potential diagnosis and treatment options that account for the patient's unique medical conditions; and improves the efficiency and effectiveness of medical care.

Email: glenn@queryhealth.ai

Website: https://www.linkedin.com/in/glennloomismd

CHAPTER 17

EXPERTISE WITH AN EXPIRATION DATE: THE HALF-LIFE OF KNOWLEDGE

By Alan Macdonald, MBA, MSc
Founder, Adaptive Lift; AI Strategist & Innovator
Toronto, Ontario, Canada

In a time of drastic change it is the learners who inherit the future.
The learned usually find themselves equipped to live in a world that
no longer exists.

—Eric Hoffer

A half-century after social philosopher Eric Hoffer issued this warning, his words ring true today. Picture a top computer engineer from the 1990s suddenly teleported into today's tech landscape; their once-cutting-edge skills in DOS and Flash programming now about as relevant as hieroglyphics. Or consider the veteran surgeon who mastered techniques in the 1970s now confronting robotic surgeries and gene therapies unimaginable back in medical school. The world has a way of rendering expertise obsolete at a jarring pace. In short, knowledge decays and it's decaying faster than ever. Degrees hang on office walls unchanged, even as the knowledge behind them quietly fades away. This is the unsettling reality of the "half-life of knowledge," where even the proudest expertise can have a best-before date.

The Reality of Knowledge Erosion

Imagine spending years mastering a skill, only to find that a few years later, the rules of the game have changed completely. This isn't a hypothetical horror story; it's everyday life in many of today's professions. The concept of knowledge having a "half-life" was first noted in the 1930s. Back then, an engineering degree had a half-life of about 35 years, meaning it took decades for half of an engineer's learned knowledge to become outdated. By 1960, that half-life had shrunk to only ten years. And today? The decay is even more rapid.

In medicine, for example, the sheer volume of information has exploded. Medical knowledge that once took 50 years to double in size (back in 1950) was projected in 2011 to double in just 73 days by 2020. No wonder doctors joke that half of what they learned in school was out-of-date by the time they started practicing. In fact, a widely cited estimate puts the half-life of medical knowledge at roughly seven to ten years, and in fast-evolving fields like oncology or genetics it is as short as two to three years. That means a newly minted specialist could find that a significant chunk of their expertise is antiquated before they're even mid-career.

Tech professionals know this all too well. The hottest programming language or IT skill five years ago might be virtually extinct today. (When was the last time you met a Flash developer?) Software engineers live in a perpetual state of learning or obsolescence. As one analysis noted, "The amount of new technical information is doubling every two years." For a student starting a four-year technical degree, half of what they learn in their first year is outdated by their third year. By graduation day, yesterday's cutting-edge tools are often museum artifacts. In psychology, too, knowledge moves on: estimates suggest an average half-life of about seven years for psych research findings.

No field is completely immune. Lawyers see laws and regulations change at dizzying speeds in new areas like cybercrime and AI. Marketing ninjas must relearn the rules every time algorithms shift or platforms rise and fall (RIP to the social media strategies built around MySpace). Even welders and mechanics periodically retrain as materials and machines evolve. If knowledge is power, it's also perishable.

Education in an Expiring Knowledge Economy

This rapid decay of knowledge has profound implications for how we educate the next generation. We are preparing students for jobs that may not even exist yet, using information that might expire before they don their caps and gowns. It sounds absurd, like giving someone fruit that rots on the way home, but it's our reality. By one famous estimate from the World Economic Forum, 65% of children entering primary school today will end up working in completely new job types that don't even exist yet. In other words, over half of today's first-graders will eventually have job titles we've never heard of. How do you educate for a future that nebulous?

Consider a child who entered grade school this year. Their eventual career might be in quantum data analytics, AI ethics law, or designing virtual reality rehabilitation programs. These are roles that sound like science fiction today. Traditional education can hardly point them toward specific knowledge for these fields because they haven't been invented. The best it can do is teach them *how to learn*. This is a dramatic shift from the days when you could go to school, learn a trade or profession, and comfortably live off that static knowledge for a lifetime. Today's students will likely need to refresh their skill sets continuously. Lifelong learning isn't a luxury; it's a necessity for staying relevant.

The data on learning trends today is both inspiring and a little scary. Here's one striking statistic making the rounds in education circles: "For students starting a four-year technical degree, half of what they learn by their first year will be outdated by their third year". Universities are grappling with this by updating curricula more frequently, but bureaucratic lag and the sheer pace of change mean many courses can't keep up. A textbook written five years ago might be stale (while costing a fortune) in fields like software engineering, biotech, or digital marketing.

Yet, there's hope in the adaptability of young minds and new approaches to learning. Educators are increasingly emphasizing learning how to learn, critical thinking, and adaptable skills over rote facts. After all, why obsess over memorizing details that might expire next year, when you can focus on mastering the art of adapting to new information? The goal is to future-proof learners, not with specific content, but with agility and curiosity. Still, the question looms: if what we learn today might be useless tomorrow, how should we value our credentials?

Degrees That Last Forever, But That Knowledge Doesn't

A degree in computer science earned in 2000 theoretically lasts forever, but does it really? That graduate might have never studied cloud computing, smartphones, or modern AI. The very things now fundamental to the field. In a very real sense, the degree has expired even if the paper doesn't have a timestamp.

The CEO of a tech firm won't be too impressed by a 30-year-old coding certificate; they care what you can do with today's technology. As the Strada Education Foundation notes, we're entering an era where "the degree's fading signal" is apparent—a one-time education at the front end of a career no longer carries people through a 40-year working life. In the face of rapid change, "a terminal credential earned at the front end of a lifetime of work no longer carries the weight it once did".

Consider how bizarre it is that most credentials never expire. Lawyers, doctors, and accountants at least have continuing education and boards to keep them up to date (would you trust a surgeon who hasn't opened a medical journal since the '80s?). But in many fields, once you have that degree or certificate, you're "set," at least on paper, no matter how out-of-date your knowledge becomes. It's like a carton of milk that magically never gets past its sell-by date, even as the contents quietly sour.

Should we rethink static credentials? Some innovators say yes. Why not have degrees with renewal requirements or digital certificates that update to reflect new skills acquired? Why not issue micro-credentials for specific up-to-date competencies rather than one macro credential earned decades ago? These ideas are gaining traction as industries realize a one-and-done education is mismatched to a world of constant flux. In fact, employers themselves are starting to shift focus from formal credentials to demonstrated skills. Surveys of hiring managers show that skills and experience often matter more than the name of a degree. The shrinkage of skill shelf-life is pushing companies to ask: what can you do today, not what did you learn ages ago?

AI: The New Catalyst for Continuous Learning

Enter artificial intelligence, simultaneously a culprit in speeding up knowledge obsolescence *and* a promising solution for keeping up with it. On one hand, AI is driving rapid change across industries, effectively shortening the half-life of skills. A recent survey of executives found that nearly half of the skills in today's workforce won't be relevant just two years from now, thanks to AI-driven changes. That's a breathtaking rate of churn. AI automates tasks, changes how jobs are done, and demands new skills (who knew "prompt engineering" would be a job a few years ago?). If you feel like the ground is shifting under your feet faster than ever, AI is one big reason.

But AI is also perhaps the best tool we have to fight the very problem it creates. How do you stay current when the knowledge landscape changes daily? You learn *continuously* and AI can help you do that at scale. Think of AI as an infinitely patient, on-demand tutor that is available 24/7. One technologist recently shared how his five-year-old son chatted with ChatGPT's voice mode for 45 minutes, going from a simple question about how cars are made into an extended learning journey. The AI adapted answers to the child's level, responded to his curiosity, and even turned a counting lesson into a fun game. As the father put it, it was like having "an essentially free, infinitely patient, super-genius teacher" that tailored itself to his kid's pace and interests. If AI can do that for a child's questions about cars, imagine what it can do for an adult trying to learn the latest programming framework or marketing strategy.

AI-driven learning platforms are emerging that can update curricula in real time. Instead of static textbooks, you might learn from interactive courses that evolve each time you access them, incorporating the latest information. For instance, if there's a breakthrough in solar panel technology today, an AI-based engineering course tomorrow morning could integrate that news into its material or quizzes. AI tutors can also identify your knowledge gaps and fill them on the fly (all by getting you to take a pre-test to assess your knowledge on a given topic). Contrast this with traditional education or corporate training, which might update on a yearly cycle at best. AI can compress that cycle to days or hours. The result is learning that's more synchronized with the pace of change.

We might even see dynamic credentials powered by AI. Envision a system where, instead of a one-time exam for a certification, you have an AI mentor/assessor that continuously checks your proficiency. This AI could pose problems, ask you to demonstrate skills regularly, and certify you in an ongoing manner. In such a world, your "credential" is not a static line on your résumé but a living profile of skills that updates as you learn (or even degrades if you don't stay fresh). Some futurists predict that "credentials that once symbolized lifelong expertise" will give way to "dynamic, ongoing assessments" of a person's actual current ability. Your expertise would be proven by continuous performance, perhaps through "lifelong conversations" with AI that verify and *augment* your knowledge in real time.

AI as a personal coach can also democratize learning. Traditionally, access to cutting-edge knowledge or elite tutors was limited to those in top universities or companies. Now an ambitious teen in a remote town can use the same AI tutor as a Silicon Valley engineer to learn about quantum computing or anything under the sun. This could level the playing field and if embraced properly, give more people the chance to keep their skills up to date. However, it also introduces a new challenge: a world where learning is turbocharged by AI will separate those who embrace this new mode from those who don't.

The New Divide: Adapt or Fall Behind

As AI-powered learning becomes mainstream, a new divide is emerging in society. It's no longer just about who has access to education, but about who has the mindset and tools for constant re-education. Those who embrace adaptive learning, leveraging AI and other resources to continuously update their skills, will surge ahead. They'll be the Hoffer-style "learners" inheriting the future. Meanwhile, those who cling to a one-and-done mindset, who stop actively learning after formal schooling will risk finding themselves left behind, equipped for a world that no longer exists.

AI itself could exacerbate this divide if not handled wisely. Those with access to AI tutors and the skills to use them will accelerate ahead. Those without access, or who lack the digital literacy to take advantage, might be stuck with an outdated playbook. It's a scenario where even within the same field, two professionals can diverge rapidly: one constantly improving via AI-guided learning, the other static. Society will need to ensure broad access to AI tools and promote a culture of lifelong learning to avoid this schism. Otherwise, we risk a future where the "knowledge rich" get richer and the rest struggle to catch up in an accelerating race.

The Evolution of Expertise: A Glimpse Ahead

Ultimately, grappling with the half-life of knowledge means embracing a simple but radical idea: education has no end point.

The mortarboard ceremony at graduation isn't the finish line; it's just one stop in a lifelong marathon. In tomorrow's world, you'll need to run that race with endurance, curiosity, and humility. The reward is not just staying relevant in your career, but the richer experience of continually growing and reinventing yourself.

For an example of how this can look in practice, consider a platform like AdaptiveLift (www.adaptivelift.ai). It's an emerging tool designed to help learners "learn anything" by starting at their current level. Through a short pre-test or quiz, it figures out precisely where you stand, then dynamically adapts the content that follows. Instead of a static course, you get real-time check-ins from quizzes, debates, games, and storytelling sessions. These challenge you right at the edge of your ability. Need an extra nudge in data analysis? AdaptiveLift offers deeper drills or case-based lessons. Ready to move faster through a familiar module? It accelerates the pace. While it's not a silver bullet, platforms like this offer a tangible glimpse of what learning in the AI era can be: hyper-personalized, evolving, and deeply engaging.

In a fast-forward future, the real expert won't be the one who learned something once and nailed it to the wall; it'll be the one who can continuously absorb, adapt, and apply new knowledge. When we embrace tools and mindsets designed for perpetual reinvention rather than static credentials, we transform the half-life of knowledge from a looming threat into an ongoing invitation to grow. If we can meet that invitation, AI stands ready to help us shape a brighter learning future: where no skill ever goes stale, no credential lives unchanged, and where the very notion of "being educated" is superseded by the art of always becoming.

About the Author

Alan Macdonald's journey begins in Vereeniging, South Africa, where a formative military stint instilled in him resilience and a profound sense of duty. While his late father remains a bedrock of inspiration, an unassuming man of deep integrity, Alan's mother is happily still with him, carrying forward the quiet strength and steady support that have shaped his life's trajectory.

Today, Alan resides in Toronto, Ontario, where he channels his restless curiosity into everything from fine-tuning barbecue techniques on his GMG pellet smoker to exploring emergent frontiers in AI and data analytics. He has earned (since 2021) both an MBA and an MSc as testaments to his commitment to staying on the cusp of evolving knowledge. Now, in pursuit of a DBA, Alan embodies the principle that learning is never truly complete. Whether he's designing AI labs like QuShift AI, co-creating adaptive learning platforms such as AdaptiveLift, or simply revisiting foundational theories, Alan is driven by an insatiable hunger to push boundaries and remain at the cutting edge of innovation.

His passion for growth is something he's proud to share with his daughter, who is also immersed in her own academic journey. Although they're in different programs, they compare notes, cheer each other on, and relish the excitement of discovering new ideas at the same time. In a world where knowledge evolves at breakneck speed, this joint pursuit of education underscores Alan's belief that staying curious and encouraging others to do the same is the surest path to staying relevant, fulfilled, and connected.

Anchored by his parents' enduring values of empathy and integrity, Alan approaches each new challenge with both humility and ambition. In a world where knowledge expires at record speed, he stands for perpetual reinvention, urging others to question assumptions and explore the unknown. Whether in business, technology, or personal growth, Alan MacDonald's driving mission is clear: stay curious, stay open, and always keep learning.

Website: https://www.adaptivelift.ai/

LinkedIn: https://www.linkedin.com/in/alanmacdonald/

CHAPTER 18

LEADING WITH VISION AND CONFIDENCE: NAVIGATING THE AI-DRIVEN FUTURE

By Tom Mawhinney, MBA, ICD.D
Author: *The Contemporary Leader*
Calgary, Alberta, Canada

> AI *won't replace humans, but those who use AI will replace those who don't.*
>
> —Garry Kasparov, chess grandmaster and AI advocate

With Change Comes Opportunity

Emma, the CEO of a mid-sized retail chain, faced declining sales and rising competition from e-commerce giants. Instead of seeing technology as a threat, Emma chose to embrace it. She partnered with an AI solutions provider to investigate the potential of embedding predictive analytics within her supply chain and AI-driven personalization within her customer engagement strategies. This ultimately led to the development of AI algorithms that analyzed purchase patterns and adjusted inventory orders automatically, reducing overstock and shortages. Simultaneously, the introduction of AI-driven chatbots provided customers with personalized shopping recommendations. Over two years, Emma's company saw a 25% increase in sales and reduced inventory costs by 15%. By viewing technology as an enabler rather than a disruptor, Emma not only reversed her company's decline but positioned it as a leader in retail innovation.

The rapid advancement of artificial intelligence (AI) and emerging technologies is reshaping the very foundations of leadership. Where once the role of a leader was primarily to oversee operations, manage teams, and ensure organizational goals were met, the integration of AI and related technologies has redefined what it means to lead. Leaders are now required to navigate uncharted waters where algorithms, automation, and data-driven insights play a pivotal role in decision-making. This transformation demands a new kind of leader—one who understands the technology not just as a tool but as a game-changer capable of enhancing human potential and organizational success. The leaders who recognize this shift early and adapt to it will be at the forefront of innovation while those who resist change risk becoming obsolete.

History has shown that leadership evolves alongside societal and technological advancements. However, what sets today's changes apart is the unprecedented speed and scale at which they are occurring. Leaders who choose to embrace these developments position themselves and their organizations to thrive. This requires a mindset shift—from seeing technology as a threat or a disruption to viewing it as an enabler of growth and efficiency. Leaders who are willing to upskill, learn about AI's capabilities, and integrate technology into their strategies will foster more resilient, agile, and competitive organizations. Conversely, those who cling to traditional methods may find themselves struggling to maintain relevance in an environment where adaptability is the key to survival.

The opportunities for leaders willing to capitalize on these possibilities are nothing short of phenomenal. AI and emerging technologies offer the chance to streamline processes, enhance creativity, and unlock new avenues of growth. For instance, AI can provide leaders with unprecedented insights into consumer behavior, market trends, and organizational performance, enabling more informed and impactful decision-making. Additionally, automation can free leaders and their teams from repetitive tasks, allowing them to focus on innovation and strategic priorities. Leaders who actively seek out these opportunities can redefine the competitive landscape, not just for their organizations but for entire industries.

At its core, leading in an AI-driven world is about balancing human ingenuity with technological advancement. While technology provides the tools, it is the human element—vision, empathy, and creativity—that gives leadership its transformative power. Leaders who harness this synergy will be the ones to shape the future, turning uncertainty into opportunity and challenge into triumph. The question is not whether change will come—it already has—but whether today's leaders are ready to seize the opportunities that accompany it.

In a world being rapidly reshaped by technology, the future belongs to those who lead with confidence, vision, and a willingness to evolve. This is not merely a challenge to overcome but an invitation to redefine leadership for the better. By embracing change, leaders can position themselves to not just survive but thrive, leaving a legacy of growth, innovation, and progress.

Fear Less, Learn More

David, a senior manager in a healthcare organization, initially felt intimidated by AI's rapid integration into his field. He feared that his lack of technical expertise would hinder his ability to lead effectively. Instead of avoiding the issue, David enrolled in a foundational AI course designed for non-technical leaders. Armed with a basic understanding of AI capabilities and limitations, David initiated a project to use AI for patient scheduling. The AI-based solution analyzed appointment patterns to minimize no-shows and optimized doctor availability. The initiative increased scheduling efficiency by 30%, reducing wait times and improving patient satisfaction. By focusing on learning, David overcame his apprehensions and transformed a challenge into a success story.

In the face of rapid technological change, it is all too easy for leaders to fall into the trap of paralysis—either overwhelmed by the sheer pace of innovation or bogged down in the endless cycle of analysis. The fear of making the wrong decision can lead to inaction, but in a world driven by AI and emerging technologies, hesitation is more detrimental than calculated risk-taking. Leadership in this era demands courage—the willingness to move forward even when all the answers aren't yet clear. Embracing an experimental mindset, where small, iterative actions are valued over perfection, allows leaders to adapt and respond in real time. To fear less is not to disregard the risks, but to recognize that the cost of standing still far outweighs the cost of calculated movement.

A critical component of overcoming fear is accepting that continuous learning is non-negotiable. The era of technological disruption demands leaders who are willing to go beyond their comfort zones and actively engage with emerging technologies. This doesn't mean every leader must become an AI expert or a data scientist. Instead, it requires an open mindset—a willingness to learn the fundamentals, ask informed questions, and appreciate the implications of these advancements. Leaders who invest in understanding how AI and related technologies work, their potential applications, and their limitations are better equipped to harness their power. Ignorance may have been tolerable in the past, but in today's interconnected, technology-driven landscape, it is a liability no leader can afford.

For leaders, the goal isn't to master the intricate details of every technological development but to know enough to form a basis for sound strategic decisions. A leader doesn't need to write the code for an AI algorithm, but they must grasp how the technology aligns with organizational objectives and the broader market landscape. This foundational knowledge allows leaders to communicate effectively with technical experts, evaluate technological opportunities and risks, and determine the best course of action for their organizations. It bridges the gap between technical teams and business strategy, ensuring that decisions are both innovative and aligned with organizational goals.

By learning more, leaders gain confidence—not just in their ability to understand technology but in their capacity to guide their organizations through its implementation. Knowledge replaces fear with curiosity, enabling leaders to approach challenges as opportunities for growth. It empowers them to lead conversations about technology's role in their industries, to inspire their teams to embrace change, and to build a culture where innovation thrives. In short, the more leaders know, the better equipped they are to lead with clarity, vision, and decisiveness.

The message is clear: fear less, learn more. The future belongs to leaders who are willing to overcome their apprehensions, invest in their own learning, and approach technology with an open mind. By doing so, they not only position themselves as adaptable and forward-thinking but also set the stage for their organizations to excel in a rapidly evolving world.

Focus on the "Why," Not the "Why Not"

Sarah, the operations head of a manufacturing firm, faced resistance from her team when she proposed adopting AI-driven robotics. Concerns ranged from job losses to skepticism about ROI. Instead of focusing on these "why nots," Sarah led with a clear "why." She explained that integrating AI could improve workplace safety by automating hazardous tasks and increase production efficiency by reducing errors. Sarah began with a pilot program, introducing robots to handle repetitive tasks on one production line. Over time, this led to a 40% increase in output and fewer workplace injuries. Sarah then reinvested the savings in upskilling programs, enabling her team to take on more strategic roles. Her focus on the benefits rather than the obstacles helped align her team and position the company for sustained growth.

As AI continues to reshape industries and redefine traditional business models, leaders must adopt a mindset focused on potential rather than resistance. The pros of AI-centric strategies—enhanced efficiency, deeper insights, personalized customer experiences, and streamlined decision-making—far outweigh the cons. While no innovation is without its challenges, the benefits of leveraging AI are too substantial to ignore. For forward-thinking leaders, the focus should be on understanding how AI can unlock opportunities rather than dwelling on the obstacles it may present. By prioritizing the "why" behind adopting AI-centric strategies, leaders can align their vision with the transformative potential of these technologies, paving the way for long-term success.

Attempting to resist the inevitable march of technology is a futile endeavor. History has shown that clinging to outdated practices and resisting change often leads to stagnation and, ultimately, irrelevance. Instead of spending valuable time and energy questioning whether AI is here to stay, leaders must accept its permanence and adapt. This shift in perspective enables organizations to move from a reactive stance to a proactive one, positioning themselves to lead rather than follow in an AI-driven world. Embracing the inevitability of change is not a sign of defeat but a recognition of the immense possibilities that await those willing to evolve. Leaders who accept this reality free themselves to focus on building strategies that harness AI's strengths.

Understanding AI is the first step; leveraging that understanding to prioritize and commit to critical initiatives is where real impact is made. AI is a tool—not a solution in itself—and its effectiveness depends on how it is applied to address organizational goals and challenges. Leaders must identify the areas where AI can deliver the greatest value, whether in optimizing operations, improving customer engagement, or driving innovation. This requires clear priorities and a willingness to commit resources to initiatives that align with the organization's strategic objectives. By staying focused on the "why," leaders can ensure that their AI investments are purposeful and impactful, driving measurable outcomes rather than being swayed by trends or hype.

Focusing on the "why" over the "why not" also empowers leaders to cultivate a culture of innovation and resilience. When teams see leadership championing AI as an enabler of growth, they are more likely to embrace it themselves. This fosters an environment where experimentation is encouraged, learning is continuous, and resistance to change is minimized. Leaders who clearly articulate the purpose and potential of AI inspire confidence, aligning their teams around a shared vision of progress and success.

In a rapidly evolving technological landscape, hesitation and doubt are luxuries that leaders cannot afford. The "why not" arguments—while understandable—should not overshadow the immense opportunities that come with embracing AI. By focusing on the "why," leaders position themselves and their organizations to thrive in an AI-driven future, turning potential challenges into stepping stones for growth and innovation.

Bring Your Team Along for the Ride

John, the CTO of a financial services firm, recognized the potential of AI in fraud detection but understood that his team was wary of the change. To address this, John hosted workshops where he introduced AI's capabilities and involved employees in the decision-making process. He encouraged team members to identify areas where AI could assist their workflows, creating a sense of ownership. When the AI system launched, it reduced fraud detection times by 50%, but more importantly, employees were enthusiastic about its success. John's collaborative approach ensured that his team felt empowered rather than displaced, transforming them into advocates for innovation within the organization.

Leadership in the AI-driven era is not a solo journey—it is a collective endeavor that requires every member of the organization to move forward together. To succeed, leaders must inspire their teams to embrace the same AI mindset they themselves have cultivated. This begins with setting the tone: demonstrating enthusiasm, curiosity, and confidence in the transformative potential of AI. Leaders who visibly champion innovation create a ripple effect, encouraging their teams to explore new ideas and adopt a mindset of adaptability and growth. Inspiration starts at the top, and leaders who bring their teams along for the ride ensure that the organization advances as a unified force.

One of the most critical roles a leader can play is helping their teams understand the possibilities AI brings, rather than focusing on perceived consequences or fears. Change often sparks uncertainty, and AI is no exception—many fear job displacement, loss of human oversight, or the devaluation of traditional skills. Leaders must address these concerns head-on, not by dismissing them but by reframing the narrative. Emphasize how AI can enhance their work, provide new opportunities for growth, and eliminate mundane tasks to make room for creativity and strategic contributions. When teams understand the "why" behind AI adoption and see how it benefits them personally and professionally, they are more likely to embrace change with enthusiasm.

Ultimately, people will only follow leaders who exude confidence and demonstrate genuine commitment to the path ahead. Teams look to their leaders for reassurance, especially in times of transformation. A leader who hesitates or projects uncertainty can inadvertently amplify doubt and resistance within their teams. Conversely, a leader who is resolute in their vision and clear about their intentions inspires trust and loyalty. This confidence must be paired with authenticity—teams can sense when leaders are merely paying lip service to an initiative. Genuine commitment, demonstrated through actions and consistent messaging, creates alignment and ensures that everyone is working toward the same goals.

Bringing your team along for the ride also means empowering them to contribute to the organization's AI journey. Encourage participation by providing training, fostering open dialogue, and recognizing the value of diverse perspectives. When teams feel included and invested in the process, they are more likely to take ownership of

AI-driven initiatives. This collaborative approach not only accelerates adoption but also strengthens the organization's collective ability to innovate and adapt.

In an AI-driven future, leadership is not just about understanding the technology—it is about inspiring people to move forward together. By instilling confidence, reframing challenges as opportunities, and fostering a culture of collaboration, leaders can ensure that their teams are not merely passengers on the AI journey but active participants in shaping its direction. Success in this era will not come from individual brilliance but from the collective power of a unified, inspired team.

About the Author

Tom Mawhinney is a distinguished board member, executive, consultant, and entrepreneur with a lifelong passion for strategy, technology, and executive coaching. Throughout his dynamic career, he has worked with organizations of all sizes, leveraging these critical skill sets to foster sustainable growth in some of the world's most competitive industries.

Recognizing the transformative power of emerging technologies, Tom has dedicated himself to staying at the forefront of innovation. His deep knowledge of artificial intelligence, robotics, and quantum computing is seamlessly integrated with his extensive commercial and corporate expertise, enabling him to provide forward-thinking guidance to leaders navigating today's fast-evolving business landscape.

As an author, speaker, and thought leader, Tom is committed to empowering others through his insights. In 2025, he published *The Contemporary Leader*, a groundbreaking exploration of the evolving skill sets required for leadership success in an era defined by technological disruption and a corresponding talent revolution. The book highlights the critical importance of technology awareness, strategic adaptability, and a human-centered approach to thriving in a rapidly changing world.

Tom's work continues to inspire leaders to embrace innovation, drive meaningful change, and lead with confidence in the face of unprecedented challenges and opportunities.

Email: tom@thecontemporaryleader.com

Website: www.thecontemporaryleader.com

AGILE LEADERSHIP IN THE AGE OF AI: BALANCING HUMAN POTENTIAL AND TECHNOLOGY IN MODERN TEAMS

By Michael Mey
Managing Consultant, AI Strategy, Agile Leadership
Zurich, Switzerland

The greatest power we have is the power to change our minds.

—Seneca

In today's organizations, the rapid advancement of artificial intelligence is fundamentally reshaping how teams work, collaborate, and create value. This transformation touches every facet of organizational life, but none more profoundly than leadership itself. As AI tools and capabilities mature, leaders are increasingly faced with a critical challenge: how can they guide their teams effectively in this new era, integrating AI in ways that enhance human potential rather than overshadow it? Beyond this, leaders must also confront a deeper and more introspective question: how must leadership culture and self-perception adapt to meet the demands of this rapidly changing landscape?

The evolution of leadership—particularly through the lens of Agile principles—offers crucial guidance for navigating this transformation. Agile leadership emphasizes adaptability, collaboration, and continuous learning—qualities that modern teams need to thrive amid the challenges and opportunities AI presents. However, this evolution also demands a cultural shift: leadership must move from being directive to being enabling, from controlling to empowering, and from static to dynamic. AI, when paired with an Agile mindset, has the potential to amplify these shifts, allowing leaders to foster innovation, drive performance, and empower their teams in unprecedented ways.

In this context, leadership cannot simply be a passive participant in the transformation; it must take the driver's seat. Leaders must assume ownership of change management, shaping not only how modern teams adopt and integrate AI but also the cultural and organizational shifts required to thrive in this new era. Agile leadership, combined with a deliberate focus on fostering a learning culture, becomes the cornerstone of navigating this transformation successfully. Only by embracing this dual responsibility—guiding both technology adoption and cultural evolution—can leaders truly unlock the full potential of their organizations in the age of AI.

Understanding Leadership Evolution

The evolution from traditional to Agile leadership marks a profound shift in how organizations manage and guide their teams. Traditional leadership, which dominated much of the 20th century, was built on principles suited to an era defined by stability, predictability, and control. Hierarchical structures ensured that decision-making flowed from the top down, with leaders maintaining control through formal processes and systems. Success was measured by strict adherence to detailed plans, and team members were often viewed as resources to be managed, with a focus on efficiency, standardization, and compliance.

However, as the pace of change accelerated and complexity increased, the limitations of this model became increasingly evident. Organizations began to recognize that rigid structures and top-down decision-making were ill-equipped to address the dynamic challenges of the modern world. Agile leadership emerged as a response, offering a fundamentally different approach to leadership and organizational culture.

In Agile leadership, the role of the leader shifts from being a director to a facilitator. Instead of commanding from above, leaders empower their teams by enabling decision-making at the level where the most relevant information resides. Teams are granted the autonomy to self-organize around objectives, fostering creativity and adaptability. Success is no longer measured by adherence to rigid plans but by the ability to deliver value and respond to customer needs with agility and precision.

This transformation represents more than just a change in management tactics—it reflects a deeper understanding of how organizations create value in uncertain and complex environments. Agile leaders recognize that solutions to today's challenges are often the result of collaborative effort, experimentation, and iterative learning rather than top-down directives. Their primary role shifts to creating an environment where teams can perform at their best. This involves removing obstacles, fostering psychological safety, and building a culture of trust that encourages open communication and transparency.

At the heart of this evolution lies the concept of servant leadership, which challenges traditional notions of authority. Agile leaders prioritize enabling others to succeed, placing the needs of their teams above their own. They embrace continuous learning, encourage experimentation, and actively support their teams in navigating uncertainty. This shift requires leaders to adapt not only their strategies but also their mindsets, moving from a culture of control to one of empowerment and collaboration. As organizations continue to face rapid technological advancements, including the rise of AI, this evolution in leadership becomes even more crucial. Agile leaders must not only respond to these changes but actively guide their organizations through them, ensuring that human potential and innovation remain at the forefront.

Leading in the Age of AI: An Agile Perspective

The integration of AI into organizational processes fundamentally changes how leaders guide and support their teams, particularly within an Agile context. This transformation creates new dynamics in decision-making, team collaboration, and the preservation of Agile principles.

AI tools provide unprecedented access to data analysis and predictive insights, transforming how decisions are made and validated. Leaders must now balance data-driven recommendations with human experience and intuition, while helping their teams maintain agility in the face of AI's sometimes rigid predictions. The key lies not in choosing between human judgment and AI insights, but in creating frameworks where both enhance each other.

Traditional Agile principles take on new dimensions in this AI-enhanced environment. The principle of prioritizing individuals and interactions becomes crucial as AI tools streamline communication and workflows. Leaders must actively preserve meaningful human interaction, ensuring that AI enhances rather than replaces the collaborative spirit that drives Agile success. This might include creating dedicated spaces for face-to-face discussions and team building, while leveraging AI to make these interactions more productive and focused.

The transformation of work patterns requires careful attention to team dynamics and engagement. As AI handles routine tasks, leaders can focus more on coaching, development, and strategic initiatives. However, this shift demands new approaches to maintaining team autonomy and creativity. Leaders must help their teams view AI as a complement to human capabilities, not a replacement for human judgment and innovation.

Customer collaboration also evolves in this context. While AI can provide valuable insights into customer behavior and preferences, it should enhance rather than replace direct customer interaction. Leaders need to help their teams use AI to inform and improve customer engagement while maintaining the human connection that drives true understanding of customer needs.

Building trust becomes more nuanced, requiring leaders to foster confidence in both AI tools and human capabilities. This includes developing team members' ability to critically evaluate AI recommendations, understanding both the potential and limitations of AI tools. Leaders must create an environment where questioning AI outputs is encouraged and where human expertise remains valued. Success in this new landscape comes from:

- Maintaining agility while leveraging AI's predictive capabilities
- Fostering human connections while using AI to enhance collaboration
- Balancing rapid decision-making with thoughtful consideration
- Creating frameworks for ethical AI use that align with Agile values
- Developing teams that can confidently work alongside AI while maintaining their creative and strategic capabilities

The role of the Agile leader evolves to include being an interpreter between AI capabilities and human needs. This involves helping teams understand when to rely on AI insights and when to trust human judgment, always ensuring that technology serves the team's goals rather than dictating them.

Building and Implementing AI-Ready Agile Teams

The successful integration of AI into Agile teams requires both thoughtful team development and systematic implementation. This dual focus ensures teams can adapt while maintaining their agility and effectiveness.

Start with a comprehensive readiness assessment that evaluates both technical capabilities and human factors. This includes understanding current team skills, existing processes, and potential areas where AI could add immediate value. Take time to gauge team members' attitudes toward AI, acknowledging their concerns and aspirations. This initial assessment creates a foundation for targeted development and helps identify meaningful starting points for implementation.

Creating psychological safety becomes paramount in an AI-enhanced environment. Team members need to feel secure expressing concerns about AI tools, admitting struggles with new technologies, and questioning AI-generated recommendations. Leaders must demonstrate vulnerability in their own learning journey while showing appreciation for team members who raise challenging questions.

Begin implementation with small, well-defined pilots that offer visible benefits. Select projects where AI can address specific pain points or enhance existing processes without major disruption. For example, start with AI-powered code review tools or automated testing before advancing to more complex applications like AI-assisted planning. This approach builds team confidence while minimizing risk.

Develop clear governance structures that address both technical and ethical considerations. Define guidelines for tool selection, data handling, and decision-making processes that combine AI insights with human judgment. These frameworks should align with Agile values while establishing clear boundaries for AI use in performance evaluation and team management.

Skills development must be systematic and ongoing. Beyond technical training, teams need to develop critical thinking skills for evaluating AI outputs and recognizing potential biases. Create regular opportunities for knowledge sharing, allowing team members to exchange experiences and insights about effective AI utilization. This prevents knowledge silos and ensures collective growth in AI competency.

Team composition may evolve as AI capabilities expand. Some members might naturally become AI specialists while others focus on areas where human skills remain crucial. Guide this evolution carefully, using cross-training and role rotation to maintain team flexibility while building diverse AI expertise across the team.

Integration into daily work requires careful attention to collaboration patterns. Help teams establish workflows that combine AI efficiency with human creativity, defining clear protocols for when to use AI tools versus when to rely on human discussion. Teams need explicit agreement on how AI tools support rather than dominate their Agile practices.

Monitor progress through both quantitative and qualitative measures. Track standard Agile metrics alongside AI-specific indicators such as tool usage patterns and quality improvements. Regular team discussions can provide insight into the human experience of working with AI tools and identify areas for adjustment.

Address challenges proactively by establishing clear feedback channels. When team members express concerns about AI tools or implementation approaches, use these conversations as opportunities to refine integration strategies and strengthen team alignment. Maintain focus on continuous improvement, adjusting approaches based on team feedback while progressing toward implementation goals.

Future Considerations

The future of Agile leadership lies in maintaining a delicate balance between technological advancement and human potential. While the evolution of AI continues to accelerate, several key areas require careful attention from leaders committed to preserving this equilibrium.

As automation of routine tasks accelerates, leaders face the challenge of redirecting human potential toward higher-value activities. This isn't simply about replacing tasks but about reimagining how teams work in an AI-enhanced environment. Leaders should focus on developing capabilities where human skills remain irreplaceable:

creative problem-solving, complex decision-making, and interpersonal collaboration.

Data literacy and human judgment must evolve in tandem. As AI tools generate increasingly sophisticated insights, teams need both the technical ability to understand these insights and the wisdom to question them appropriately. Leaders should foster an environment where analytical capabilities and human intuition complement each other, ensuring AI tools enhance rather than replace critical thinking.

The ethical dimensions of AI implementation require a balanced approach. Leaders must ensure their AI strategies maintain fairness and transparency while protecting team autonomy and professional growth. This includes preventing algorithmic bias, establishing clear boundaries for AI use, and considering the broader implications of AI adoption. Creating frameworks for ethical decision-making helps teams address these concerns proactively, ensuring that efficiency gains don't come at the cost of human well-being and development.

Conclusion

The integration of AI into Agile environments presents both opportunities and challenges for leaders and their teams. Success depends not on wholesale AI adoption, but on thoughtful integration that preserves and enhances the human elements that make Agile methodologies effective.

Effective Agile leadership in the AI age requires maintaining focus on human connections while viewing AI as a tool to augment human capabilities. Leaders must preserve Agile principles while leveraging new technologies, building team confidence in working alongside AI while creating environments where both people and technology thrive.

The path forward involves continuous learning, adaptation, and commitment to human-centric values. By approaching AI integration with intention and care, leaders can help their teams harness new technologies while strengthening the collaborative, adaptive spirit that drives successful Agile organizations. The key lies in remembering that technology serves the team, not the other way around.

About the Author

Michael Mey brings over two decades of experience to the intersection of AI, requirements engineering, and Agile methodologies. His journey began with a diploma in computer science (Dipl.-Inform. FH) from Hochschule Furtwangen University, where he first worked with neural networks and artificial intelligence.

His professional path spans major financial institutions, including a leading Swiss bank, global consulting firms, and European stock exchanges, where he specialized in bridging business needs and technical implementation through requirements engineering practices.

In 2018, Michael co-founded Obvious Works, a consultancy focused on understanding client needs through advanced requirements engineering, implementing Agile leadership practices, and developing practical AI strategies. His expertise in these three critical domains allows him to help organizations create solutions that address business needs while embracing technological innovation.

Email: michael.mey@obviousworks.ch

Website: https://obviousworks.ch

LinkedIn: https://www.linkedin.com/in/michaelmey/

CHAPTER 20

THE BLURRING FRONTIER: AI AND THE EVOLUTION OF REAL AND VIRTUAL WORLDS

By Damien Montessuit
CEO, Strategic Advisor, Board Member
Lyon, France

The real voyage of discovery consists not in seeking new landscapes,
but in having new eyes.
—Marcel Proust

In the not-so-distant future, a dawn breaks unlike any other before it. It is a dawn not of the physical sun rising in the east, but of artificial intelligence illuminating the vast potential of the virtual world. Just as the rays of the sun touch every corner of our planet, AI's reach extends to every corner of human civilization, enabling individuals from all walks of life to craft and cultivate vibrant virtual landscapes in parallel to their real-world existence.

The Craftsmanship of Dreams

Imagine an artisan in a small village in Provence, who, with the touch of a finger, transforms their handcrafted pottery into digital masterpieces admired by art lovers worldwide. In the heart of bustling Tokyo, a student perfects their knowledge of ancient civilizations through immersive AI-powered simulations, each detail more vivid and accurate than the last. Across the globe, every person is not just a passive observer but an active creator in this vast digital realm.

This convergence of AI and virtual worlds is reshaping our understanding of creativity and craftsmanship. Traditional skills are being augmented and transformed by digital tools, allowing artisans to reach global audiences and preserve cultural heritage in ways previously unimaginable. The virtual space becomes a canvas for unlimited expression, where the constraints of physical materials no longer apply.

Parallel Lives

In this new era, the virtual world is not a mere escape but an extension of reality. It is a place where dreams are molded and aspirations are achieved. Farmers digitize their fields to analyze and optimize crop yields; doctors simulate complex surgeries to save lives with unprecedented precision; architects build cities in cyberspace before a single brick is laid. Each profession, hobby, and passion finds its counterpart in this parallel universe, rich with possibilities and free from physical constraints.

This blending of real and virtual worlds is transforming how we work, learn, and interact. The concept of a "digital twin"—a virtual replica of physical assets, processes, or systems—is expanding beyond industrial applications to encompass entire lifestyles. People are creating digital versions of themselves, their environments, and their daily activities, allowing for unprecedented levels of analysis, optimization, and experimentation.

As AI technologies advance, they are revolutionizing the creation and manipulation of virtual environments. These AI-generated realities are becoming increasingly sophisticated, adaptive, and indistinguishable from physical reality. Advanced algorithms can now generate hyper realistic landscapes, characters, and entire worlds that respond dynamically to user interactions.

In education, AI-driven virtual environments create immersive learning experiences that adapt to individual learning styles and preferences. Students can explore historical events as if they were present, conduct virtual scientific experiments with real-world implications, or practice complex skills in safe, simulated environments.

In healthcare, AI-generated virtual realities are transforming patient care and medical training. Surgeons can practice complex procedures in highly realistic simulations, while patients undergoing rehabilitation can engage in tailored virtual therapy sessions that adapt to their progress in real time.

The Metaverse and Beyond

The concept of the metaverse—a shared, persistent virtual space—is rapidly evolving from science fiction to reality, largely thanks to AI. This interconnected network of virtual worlds is becoming a new frontier for social interaction, commerce, and cultural exchange. AI plays a crucial role in making these spaces feel alive and responsive, populating them with intelligent non-player characters (NPCs) and facilitating natural interactions between users.

As these virtual spaces become more sophisticated, they are not just mimicking reality; they are augmenting it. AI-powered augmented reality (AR) technologies are overlaying digital information onto our physical world, creating a hybrid reality that enhances our daily experiences. This blending of real and virtual elements is reshaping how we interact with our environment and each other.

AI Companions and Digital Humans

One of the most intriguing developments in this space is the emergence of AI-powered virtual beings. These digital entities, ranging from chatbots to fully realized "digital humans," are becoming increasingly sophisticated and capable of forming emotional connections with users. As AI advances, these virtual companions are blurring the lines between human and artificial relationships.

The rise of AI companions raises profound questions about the nature of consciousness, empathy, and human connection. As people form deep bonds with AI entities, we may need to reconsider our definitions of friendship, love, and even personhood. The psychological and social implications of these relationships are only beginning to be understood.

Surviving and Thriving

As AI reshapes our world, survival in this new landscape requires adaptability and foresight. The key to thriving lies in leveraging AI as an ally, rather than viewing it as a mere tool. Continuous learning becomes the lifeblood of progress, with individuals embracing lifelong education to keep pace with the rapid advancements. Building a robust digital literacy, understanding AI ethics, and fostering creative problem-solving skills are the new pillars of survival.

The Heartbeat of Humanity

Despite the digital transformation, the heartbeat of this brave new world remains human. Virtual worlds become a canvas for human expression, collaboration, and innovation. Here, the shared experiences in digital realms deepen our understanding and appreciation of our real-world counterparts.

The convergence of AI and virtual worlds presents both opportunities and challenges for maintaining and strengthening human connections. AI-powered virtual environments enable seamless collaboration across geographical boundaries, fostering cultural exchange and global understanding. Virtual spaces offer novel ways to connect with others, potentially reducing isolation and creating communities based on shared interests rather than physical proximity. AI technologies can enhance our ability to understand and relate to others by providing insights into emotional states and facilitating more effective communication. Virtual worlds offer unprecedented opportunities to preserve and share cultural heritage, allowing people to experience and interact with historical sites and artifacts in immersive ways.

Ethical and Philosophical Implications

The blurring of real and virtual worlds raises profound ethical and philosophical questions that society must deal with. As individuals create multiple digital identities or avatars, questions arise about the nature of personal identity and authenticity in virtual spaces. The vast amounts of personal data generated in virtual environments raise concerns about privacy, data ownership, and potential misuse of information. As virtual worlds become more integral to our lives, new frameworks for digital rights and governance will need to be developed. The increasing realism of virtual environments challenges our understanding of what constitutes "real" experience and raises questions about the nature of consciousness and perception. Ensuring that AI systems in virtual worlds are developed and used ethically, without perpetuating biases or causing harm, becomes crucial.

Economic Disruption and Opportunity

The convergence of AI and virtual worlds is reshaping economic landscapes, creating new opportunities while disrupting traditional industries. The rise of virtual economies within games and metaverse platforms is creating new forms of value and challenging traditional economic models. Emerging technologies are creating demand for new skills and professions, such as virtual architects, AI ethicists, and digital asset managers. Virtual worlds can potentially level the playing field, allowing individuals from diverse backgrounds to access global markets and opportunities. As more activities move to virtual spaces, traditional industries may face disruption and need to adapt to remain relevant.

The Continuously Changing Landscape

As we venture further into this AI-enabled future, Marcel Proust's wisdom resonates more profoundly than ever: "The real voyage of discovery consists not in seeking new landscapes, but in having new eyes." Indeed, the transformative power of AI and virtual worlds lies not merely in the digital landscapes we create but in how they fundamentally alter our perception of reality itself.

The tapestry of civilization becomes ever more intricate and beautiful, woven with threads of both physical and virtual experiences. Each thread represents not just a life, but a new way of seeing, understanding, and being in this converged reality. The boundaries blur not to confuse, but to enrich, offering a limitless playground for human ingenuity and resilience.

This convergence of AI and virtual worlds represents one of the most significant shifts in human experience since the dawn of the digital age. Like Proust's metaphorical "new eyes," these technologies give us unprecedented ways to perceive, create, and connect. Yet, as we navigate this new landscape, we must remain mindful of both its immense potential and inherent risks.

The challenge lies in shaping this convergence in ways that enhance rather than diminish our humanity. As Proust understood about the nature of discovery, our journey into this new frontier is not about the virtual worlds we build, but about the new perspectives we gain and the deeper understanding we develop about ourselves and our place in both physical and digital realms.

In this brave new world, every person becomes not just a creator or architect, but a pioneer with "new eyes" to perceive and shape their destiny across both real and virtual domains. The future is unfolding all around us, and through these new lenses of perception, we have the opportunity to craft it with wisdom, empathy, and purpose. The true discovery in this technological revolution lies not in the digital landscapes we create, but in our transformed ability to see, understand, and experience the world in entirely new ways. The question that faces us now is not just how we will adapt to this changing world, but how we will use our "new eyes" to shape it for the better.

About the Author

Damien Montessuit is a dynamic global business leader and strategic advisor with extensive executive experience across large corporations and scale-ups in diverse industries. He provides chief-as-a-service missions, driving strategic growth and transformative initiatives. Damien's leadership style blends hands-on experience with forward-thinking vision, emphasizing continuous learning and cross-cultural collaboration.

Passionate about the potential of artificial intelligence and its applications across various sectors, Damien serves as an advisor for multiple companies exploring innovative projects at the intersection of AI and business, helping organizations navigate the exciting possibilities of this transformative technology.

Damien holds a master's degree in computer science and a TRIUM Global Executive MBA (HEC, LSE, NYU), highlighting his commitment to lifelong learning and his diverse global business insights.

LinkedIn: https://www.linkedin.com/in/damienmontessuit/

CHAPTER 21

A JOURNEY BUILDING AUTHENTIC COMMUNITIES WITH INNOVATION

By Peggy O'Flaherty
Co-Founder & CEO, VINLY; AI Messaging Expert
Chicago, Illinois

Don't be intimidated by what you don't know. That can be your greatest strength and ensure that you do things differently.

—Sara Blakey, CEO of Spanx

Two years ago, while eating dinner with my teenagers, we were talking about the controversy of students using ChatGPT to write papers. The world that I knew was changing and that felt intimidating. Now, I see the promise of artificial intelligence, especially around purpose-built solutions that bring people together in a transformative way.

In a world where billions of connections are just a click away, why do so many people still feel isolated? This is something

I'm especially aware of as a mother of five teenagers. Social media promised to bring us closer, but it left us chasing "likes," scrolling endlessly, and building networks that often lack depth or purpose. In business, tools like LinkedIn are filled with automated responses and direct messages for sales opportunities. Imagine if technology could do more—not just connect us, but empower us to solve genuine problems, foster meaningful relationships, act out of compassion, and create thriving communities.

Building community is in my blood (this is unsurprising as I am the youngest of ten children). The value of community inspired me over the last 15 years to build technology for the greater good, technology that helps convene. My first effort was Growing the Faith and the OneParish app to help churches grow engagement with their members, which was acquired in 2017. My second company was Mavely, the everyday influencer platform. It enabled creators and influencers across major social channels to share their favorite brands. It too was acquired. It sold in 2021 and then again in 2024 for $250M. My current venture, Vinly, grows connections starting with 1 on 1 conversations. Then we weave AI in the background to do the heavy lifting for engagement and connections. This transforms simple interactions into engaged communities.

Churches around the country struggle with building communities. Growing the Faith's OneParish app was a tool to unite local communities around specific needs in the context of a Catholic parish. Adoption of a customized church app in 2013 was slow. We found more success, as my children took part in the launch at each church. People responded well to the five of them dressed in their matching outfits, providing instruction to older parishioners. They held their hand and encouraged them to download and discover new ways to engage with the community. The platform made it easy for community members to take part in local food pantries, donate clothing to those in need, and come together to clean up local parks. It became a communication platform to support their members in faith formation, service, and stewardship.

While I was building growth within churches, I was also working with a group of moms who were capitalizing off their friends and relatives by hosting home parties and building multilevel marketing organizations. The backlash from people in my network felt isolating. That feeling drove me to build a more authentic way for women to share their favorite brands and use technology to expand their reach. My second attempt to build communities with innovation was Mavely. Growth at Mavely came with the boom of the creator economy and social media. The world's largest brands sought out authentic women and men sharing brand affinity across social platforms.

While the growth of Mavely was beyond my wildest dreams, the isolating impact of social media and greed from consumerism also struck me. Stanford University's Social Media Lab explored the impact of social media on isolation. The researchers found that while social media can help people feel connected and reduce feelings of isolation, especially during times of physical distancing, it can also lead to negative emotional experiences if overused or used as a substitute for in-person interactions. The design of social platforms maximizes addictive engagement, resulting in superficial interactions instead of meaningful connections. Businesses and individuals find themselves overwhelmed with irrelevant content, and communities are left fragmented rather than unified. Technology should go beyond shallow exchanges and create networks rooted in purpose. I was on a quest to build again. I was in search of a tool that brought people together to create deeper synergies and collaborations to reshape how we solve bigger problems in our world.

It is human nature to want deeper connections with family, friends, and in business. In business we win when we create connections that foster win-win experiences between teams, customers, vendors, and partners. However, networking tools and events in business challenge us with noise, ads, and self-serving behavior. We can provide better service when we implement tools that extract sentiment and analyze customer preferences, behaviors, and connections. We can scale faster with tools that help acquire customers and streamline workflows. Businesses will have better margins when the tools are easy to implement and affordable. These tools exist, but from my perspective have only been developed as enterprise solutions.

In our everyday life, adoption of tools embedded with AI is happening, on our phones, appliances, and cars. But what about our communities and daily interactions with each other? Social graphing and large language models (LLMs) are now being used to analyze patterns, behaviors, and shared goals to connect people and businesses in ways that traditional platforms cannot. For small businesses, this means finding the right customers, collaborators, or mentors, not through endless advertising but through purposeful matchmaking. For communities, it means fostering relationships that strengthen local economies, bridge divides, and activate networks to solve shared challenges. AI transforms the idea of a network from something passive into something dynamic—a living, breathing system of support and innovation. Imagine if every local coffee shop could adopt a more personalized offering with a point-of-sale tool that had artificial intelligence embedded into it to recommend a seasonal drink to a regular customer based on their previous purchase history. What if every customer was as known to the employees as Norm was to the bartenders in *Cheers*?

A local restaurant, retail shop, software company, or enterprise business can engage tools like social graphing to find common threads that connect their customers at a deeper level. Connecting people based on shared location, values, goals, and opportunities can lead to better collaboration. For example, a local bakery might not just find more customers but could connect with nearby coffee shops or event planners to create mutually-beneficial partnerships. These purposeful networks turn competition into collaboration, fostering ecosystems where small businesses and communities thrive together rather than competing for limited attention. Consumers win and the businesses that serve them win.

What makes this revolutionary is its ability to go beyond data and into forged connections that matters. We will have at our fingertips a more precise way to network, better than Facebook or Linkedin by connecting individuals and businesses based on shared goals, needs, values, and geography. Instead of generic Linkedin connection requests or the endless DMs of someone trying to sell you something, imagine connections that can help you achieve your to-do list or AI agents that can help you work five times faster. A vector database goes beyond acting as a catalog of information—they interpret it, finding patterns in human behavior and intent that would otherwise go unnoticed. This means that we can uncover opportunities for collaboration on a scale we've never seen before. For instance, Annie Krug, author and nonprofit organizer, is connecting third-party logistic companies and nonprofits with the help of AI. She identifies inventory that is abandoned and headed to a landfill and rescues those items, placing them in local homeless shelters or crisis centers. These connections happen seamlessly, powered by data but rooted in shared purpose. AI doesn't just connect dots; it empowers people to work together in ways that turn ideas into action and communities into catalysts for change.

An example of that would be AstraZeneca. A global, science-led biopharmaceutical company that focuses on the discovery, development, and commercialization of prescription medicines employed graph databases to visualize patient journeys, enabling them to answer complex questions about prescriptions and diagnoses. This approach facilitated a deeper understanding of patient experiences and contributed to improved healthcare delivery. This is one example of technology that is now being implemented, making connections that can save lives and redefine how we think about healthcare.

As powerful as AI is, its potential comes with significant ethical responsibilities. The systems we build are only as good as the intentions and safeguards behind those who create them and the AI LLMs they use. Left unchecked, AI risks perpetuating bias, widening inequalities, or being used for win/lose profit-driven motives that undermine its ability to create meaningful change. Developers, businesses, and policymakers have a collective responsibility to ensure that AI remains a tool for empowerment rather than exploitation. This means prioritizing transparency, fairness, and inclusivity in AI design—ensuring that it serves everyone, not just the privileged few. Ethics must aim to connect people with purpose, amplify underrepresented voices, and prioritize solving real-world problems over chasing profits or engagement metrics. LLMs like GPT are pivotal in making AI accessible and transformative for communities and businesses alike. These models enable natural, intuitive interactions with AI, breaking down barriers of complexity. The ability of LLMs to understand and process human language at scale opens doors for unprecedented global collaboration.

It is equally as important to be mindful of AI biases. In 2019, a study conducted by MIT Media Lab researcher Joy Buolamwini and AI ethicist Timnit Gebru evaluated the performance of Amazon's facial analysis software, Rekognition, in identifying individuals' gender across different demographics. The research revealed significant disparities in accuracy. The findings indicated that Rekognition's performance was less reliable for individuals with darker skin tones and for females, suggesting inherent biases in the system. In response to these findings, Amazon contested the study's conclusions, asserting that the researchers had misinterpreted the results because of an improper "default confidence threshold." This case underscores the critical importance of addressing and mitigating biases in AI systems, especially those deployed in sensitive applications like facial recognition.

With the right ethical guardrails and technological advancements, AI has the potential to tackle some of the world's most pressing problems. For example, Google Earth Engine integrates various satellite imagery and geospatial analytical capabilities. It enables users to visualize and analyze data related to forest cover, agricultural land, surface water, and more. This helps researchers and policymakers make informed decisions about environmental conservation and resource management. The world needs solutions that help solve problems and leaders in tech who are rooted with deeper values. Nova Spivack, CEO of mindcorp.ai, wrote a post about how social media can be a channel for more compassion in our world: "I do think a more compassionate society can be engineered with the science of memetics (the study of behavior, style or culture that spreads from person to person) and spread via social media, the same way our present less compassionate society is spreading the opposite messages via social media."

Building ethical guardrails has been a priority in each company I've built and especially now with Vinly. We performed early user interviews to understand the challenges with building community. During our interviews, we heard adamant feedback on leveraging AI as a digital clone. Community builders resonated with the idea of sharing their knowledge for a bigger impact. Intrigued by the possibilities, they explored new applications of AI for community benefit. We heard a desire to reduce the noise, in their day, from excessive inbound emails, SMS, multiple channels for messaging, and a lack of overall trust from various platforms. At Vinly, we knew that building with privacy and legal compliance was our goal. We heard loud and clear that they wanted to better connect with those they know and with those who are well known and trusted by those they know. LinkedIn, Slack, and Discord are leveraging AI to connect with various teams or groups with a common theme inside or outside an organization. But, they are missing a critical element of trust: actual connections with people in our trusted networks with known intent to do good.

Trust is a core value in building communities and technology. AI systems must actively build and maintain trust by prioritizing privacy, transparency, fairness, and accountability throughout their life cycles. This begins with designing algorithms that are free from bias and audited to ensure ethical performance. Clear communication about how the AI operates, what data it uses, and the decisions it makes can help users feel confident in its capabilities. Establishing safeguards, such as explainable features and feedback loops, ensures that users understand and can question a system's outcomes. Ongoing monitoring and updates are essential to adapting to new challenges and maintaining reliability. By aligning development with human values and involving diverse perspectives in its design and implementation, these systems can foster the trust needed to support long-term adoption and meaningful engagement.

Our mission at Vinly is to weave people and AI together. We are building with the intent to enhance the quality and quantity of human interactions. We achieve this by providing a place for people to convene around shared topics and by fostering deeper interaction. AI-assisted messaging fosters authentic connections, supporting clients, guiding students, or managing a family schedule. By automating routine tasks and establishing consistent meaningful connections, Vinly empowers you to focus on what matters most: building trust, offering value, and nurturing the relationships that drive impact and success. The tool is there to help organize the thousands of good conversations and avoid the noise of other platforms. This should facilitate deeper relationships with customers, networks, and communities. With AI the context driven from LLMs gives us the sentiment analysis of conversations and leads to deeper decisions, as the AI is thinking like humans. From this comes pairing people within networks, who may never find each other otherwise.

The promise of AI goes beyond its ability to write an email, crunch data, and automate tasks. Its promise lies in its capacity to bring people together in ways that foster collaboration, innovation, and empathy. We have the tools now to shift from social networks driven by vanity and win/lose profit to systems designed for purpose and impact, that generate higher profits because the users win too. Imagine a world where every small business finds its perfect partners, every community has access to community-building tools that address resources tailored to its unique needs, and every community is powered through collective action. This future isn't just possible—it's within reach. But achieving it requires us to embrace AI responsibly, prioritize platforms that generate real connections over clicks, and focus on solving the problems that matter most. Imagine a world where people feel connected, purposeful, and empowered to share those feelings with others. The question is no longer what AI can do—it's what we will choose to do with it.

About the Author

Peggy O'Flaherty is a mother of five children, Grania, Enya, Cillian, Liadan and Orla, and a serial entrepreneur with five startups under her belt. Peggy's focus has been to build communities around their favorite brand and leverage technology for greater reach. She has had two successful exits: Growing the Faith's OneParish App and Mavely. Peggy is co-founder and CEO of Vinly and founder of POF Consulting. Peggy does consulting, advisory, and public speaking. Her core strength is customer acquisition, growth hacks, and building great culture. Peggy spends her free time baking for family and supporting nonprofits on mission trips to Nicaragua and Africa.

Email: Peggy@Vinly.co

CHAPTER 22

NAVIGATING THE LEADERSHIP LANDSCAPE: DATA VS. SOFTWARE ENGINEERING

By Ravi Pasula
Head of AI & Machine Learning
San Jose, California

During World War Two, the military wanted to reinforce aircraft based on the patterns of bullet holes found on returning planes. The seemingly logical approach was to reinforce the areas with the most bullet holes. A brilliant statistician realized that the planes *returning* were the ones that had survived despite those hits. The areas *without* bullet holes on the returning planes were likely the critical areas that, if hit, would cause the plane to be lost. Therefore, those were the areas that needed reinforcement. The statistician was Abraham Wald. However, Abraham is not as famous as other war heroes like Audie Murphy (US) or Guy Gibson (UK).

Now imagine two rising stars: one a brilliant software engineer, the other a data science prodigy. While both are destined for leadership, their paths to the top might look surprisingly different. Are you drawn to the power of data or the magic of code? Understanding the distinct growth paths of data leadership and software engineering is crucial for anyone navigating the tech landscape. As a data leader with 15-plus years' experience, I hope to use my observations in Silicon Valley and share how data employees can grow as leaders. We will review key milestones in the careers of data employees, namely: *maturity of data literacy, the intersection of the personal and professional, and technical depth versus strategic depth.*

Maturity of Data Literacy

After finishing my master's in computer science, I was thrilled to join a telecommunications billing company. My first job along with being in the land of opportunity, the USA, was a liberating experience. I was responsible for ingesting wireless messages and routing them for a wireless billing application. As I worked long hours in the data systems, I observed that the software engineering teams had clearer and faster career growth trajectories. The software engineering team deliverables, like billing websites, were easily visible to our customers and leaders. It felt like the software engineering path was well established. While I did (and still do) enjoy working with data, I observed many members of my team were worried about being stagnant in their roles.

It was not until four years later, when I joined a large web portal and search engine company, that I realized the value of having data literacy in leadership. In this company, we were handling petabytes of data. I started as an individual contributor (IC), but within two years, I was assigned to be a team manager. The company was investing towards building the world's largest search and advertising platform. The company was creating opportunities uniformly across software engineering and data organizations.

As an example, during my time there, I had an opportunity to build a brand-new program called Revenue Protection Team. The program was doing statistical analysis of incidents with advertising and data systems. In an organization where data maturity in leadership is highest, we interacted and promoted great ideas to build great data products. The mentorship we gained in these organizations is an asset for a lifetime. Some of my mentors whom I do not work for anymore still connect with me and guide me in various career situations.

Today, data roles are evolving with a burst of AI opportunities, and many companies are figuring out how to enable clean data consistently and fast. Companies with mature data literacy appreciate the value of these data engineers.

The Intersection of the Personal and Professional

At 2 a.m., an alert goes off about a production incident that could have a significant impact on revenue. I worked with my team until 4 a.m. to resolve the incident. At 7 a.m., I woke up as my four-month-old son was awake and needed my attention. By 9 a.m. I had to be in the office to review the incident's post-mortem and build a remediation plan. Indeed, that was a busy time.

Data management's always-on culture fueled by instant communication and global collaboration can make it difficult to disconnect and recharge. How events in our personal lives impact our professional journeys isn't something discussed much in my field of work. Regardless of the data or software engineering organization, work-life balance issues are often unspoken realities.

As a different example, once a co-worker asked one of my teammates if he would be willing to help with finishing a feature by staying late that evening. The co-worker even suggested that as the teammate was not married and did not have kids, it would likely be easy for him to put in the extra work. My teammate replied that he still had life and went home.

Around this time, most of my best mentors at the search and advertising company had left. I remember one of my mentors told me, "Look around at which leaders are still with you, and if none has genuine talent, you should leave too." I did leave the search and advertising company and went to a large entertainment and theme park enterprise. Some career advice: sometimes change is refreshing.

Meeting new teammates, new leaders, and new technologies triggered in me a new opportunity to start new habits in my work and with my career. While the work pressure still existed, I was getting better at managing work-life balance, and I now had two children. As a manager, I also understood why delegation is powerful. I always enjoyed staying close to the technical details but had to find the balance. I was more involved in passionate brainstorming of design/architecture and unblocking when feature delays were encountered.

After a few years at the new place, I had a feeling of career stagnation. I thought to myself, "I used to grow and expand my scope so fast in past years, why am I slowing now?" In hindsight, it was a natural progression. The slowdown in my career was preparing me for my next opportunity. I describe the experience in the following graph:

WHAT WORK

Expectation Vs Reality

How You Think Your Career Progression Should Be Like

How It Actually Is

Technical Prowess vs. Strategic Vision

While organizations play a crucial role in fostering work-life balance, ultimately, it's the individual who must navigate their career path. And that path often involves a critical decision: how to balance the development of technical skills with the cultivation of strategic thinking. This brings us to the essential distinction between technical prowess and strategic vision.

In the world of software engineering, technical debate is everlasting with great debates on which programming language is better. The arguments ranged from modern languages like Rust to traditional C/C++. The debates would center around the handling of memory safety, performance, and/or concurrency. The debates would sometimes spill over to writing test code using massive random test data and comparing the results of both languages. These healthy discussions are needed and drive motivation for engineers. Sometimes these discussions also lead to innovations by applying the right software to the right problem. This acted as a reminder that around this phase, there was a never-ending learning curve for software engineering professionals. However, as a data enthusiast, I used to wonder how business leaders can capitalize these ideas and innovations.

Historically, data's technical stack was relatively slow-changing for the most part. Traditional databases had ruled most parts of the data tech stack. SQL, NOSQL, and PL/SQL were the most common tools for data engineers, data analysts. R was the dominant data science tool. The key strengths of data teams were, however, to enable strategic visions for enterprises, which would help them to make the best business decisions. As data volumes grew, traditional databases could not manage the scale and performance. Hadoop and its equivalents transformed data teams' value propositions in organizations. Data evolved to be a key enabler for customer-facing products like advertising, recommendations engines, ride-sharing apps, and more.

Personally, I ponder what advice I would give to my younger self if I had choices on what aspects of data I should work on. My choices would be to work on large (exabyte) data systems in a

RAVI PASULA

struggling technology company in product vision or medium size (100 to 200 terabytes) data systems in a company with a strong business data acumen. I think data teams need to experience both worlds.

Getting to work on a large data system is rare, and very few companies in the world have these kinds of data. The scale and challenges of running data at massive scale make for a complicated data-engineering effort. On the flip side, understanding how to deliver business value using data of any volume is important for aspiring data engineers.

AI, AI, and AI

Innovation in data is not complete without mentioning AI. During the early part of my career, we used to do anomaly detection over large volumes of real-time data. Our methods were not called AI but were part of regular data engineering. My teams had also done attribution of mobile user behavior sessions, which would enable targeting cohorts/segments for marketing efforts. I personally had downloaded first versions of C source code for Word2Vec. Word2Vec was one of the early NLP tools to detect synonymous words or suggest additional words for a partial sentence.

This curiosity led us to building sentiment analysis for an entertainment media company's movie release trailer. While AI was still not termed as a mainstream job classification, we were developing AI-based solutions for video feeds, recommendation engines, and pricing optimization models. To continue to expand my knowledge, I used to study several data science and neural network books and courses at Coursera. As engineers wonder how they can make the switch to the next new technology (e.g., quantum someday), it's important that you are ready knowledge-wise and are working at a company that's willing to experiment with controlled experiments using the new technology.

Conclusion

In the current technology world, both data and software engineering domains are moving at an equally rapid pace. Companies are creating new roles for both functions, and both functions must keep learning to understand how technologies will influence business and transform user experiences. We are moving to a world where data, AI, and software engineering will be built as one product. To grow in these fields, you need to have a genuine passion for learning new technology and surround yourself with great mentors.

About the Author

Ravi Pasula is a data enthusiast with bachelor's and master's degrees in computer science. His career spans diverse industries, including telecom billing, online search and advertising, media, theme park entertainment, and colocation services. He has held leadership positions as manager and senior director, managing teams ranging from ten to 100 members. Proficient in handling data systems of varying volume and velocity, Ravi has a patent filed in advertising and two more in process. A dedicated mentor, he guides Army veterans transitioning to corporate careers. A voracious reader, with over 150 digital books, Ravi's passion lies in building impactful data products. He aspires to become a technology strategist or entrepreneur. Outside of work, he enjoys thriller and drama movies. Ravi credits his success to numerous mentors and the unwavering support of his family and parents.

LinkedIn: https://www.linkedin.com/in/ravi-kiran-pasula-a543404/

CHAPTER 23

BEYOND TRANSPORTATION: POWER OF AUTONOMOUS VEHICLES

By Pramod M. Patke
Product Manager, System Architect
Gothenburg, Sweden

> *The best way to predict the future is to create it.*
> —Peter Drucker

What if your commute became a moment of productivity or relaxation instead of stress? Imagine catching up on work, reading, or enjoying the scenery while your autonomous vehicle (AV) navigates the traffic. AI-powered AVs are set to transform not just transportation but also retail, healthcare, and urban planning. Yet, their full impact remains uncertain, presenting both vast opportunities and challenges.

The AV revolution calls for proactive shaping to ensure a sustainable, equitable, and connected future.

A Journey Tailored to You

Imagine stepping into a car that doesn't just take you places; it understands you. With artificial intelligence at its core, your vehicle evolves from a tool into a trusted companion, adapting to your needs, preferences, and even your emotions. This is the transformative promise of personal mobility in the era of AI.

The Empathetic Car

Advanced emotion-sensing technology detects stress, fatigue, or anxiety, adjusting the vehicle environment to create a calming and rejuvenating experience. Dimmed lights, soothing playlists, or even scenic detours can turn stressful commutes into moments of mental respite.

In 2023, a pilot program in Stockholm used AI-powered AVs that adjusted their environment based on passengers' emotions. Through biometric sensors, the vehicle detected stress and automatically adjusted seat positions, dimmed the lights, and played calming music. This illustrates how AVs could evolve to not only provide transportation but also enhance mental and physical well-being during daily commutes. Consider a future commuter Sara, who works long hours at a high-pressure job. As she steps into her AV after a tense day, the vehicle detects her heightened stress levels and gently adjusts the lighting, plays her favorite relaxation playlist, and offers her a scenic route home through a tree-lined avenue. Instead of arriving home exhausted, she steps out of the vehicle feeling more at ease, as if the car itself understood her emotional state. Such interactions redefine our relationship with transportation, transforming it from a passive tool into an active facilitator of well-being.

Productivity and Learning

Autonomous travel is reshaping the way we use our time on the road. Rather than spending minutes or hours navigating congested streets, passengers can reclaim these moments to enhance productivity, personal growth, or relaxation.

For instance, business professionals may conduct meetings, students may take virtual classes, and creatives may brainstorm ideas all while traveling. AVs may even integrate with augmented reality (AR) and virtual reality (VR), turning the cabin into a mobile workspace or entertainment hub. This shift forces us to reconsider our definition of "commuting": is it simply about getting from A to B, or can the journey itself become a fulfilling experience?

Environmental Awareness and AI as a Guardian

Beyond comfort, AVs are also evolving to become environmental stewards, using advanced perception systems to safeguard not just passengers but entire ecosystems. For example, AI-driven vehicles can detect underground vibrations, identifying potential hazards like sinkholes or road deterioration before they become visible. In rural or forested areas, they use thermal imaging and AI to detect wildlife near roads, preventing collisions and protecting ecosystems. In a world where climate consciousness is growing, these AI-powered guardians will help cities and individuals make better choices, ensuring that transportation isn't just smart, but sustainable and responsible.

Urban Development

As AVs transform personal and shared mobility, they also hold the potential to reshape the very fabric of our cities. By reducing the dominance of personal cars, AVs unlock the potential to create smarter, greener, and more people-centric environments.

Reclaiming Urban Spaces

Maya, an urban planner, walks through what was once a congested parking lot but is now a thriving green space. Where rows of cars once stood, children play in community parks, and small cafés line pedestrian-friendly streets. With AVs reducing the need for parking, cities can reclaim vast amounts of space for public use, enabling parks, affordable housing, and recreational areas to flourish.

Smarter Infrastructure

AVs demand precision, not excess. Roads could become narrower, intersections smarter, and lanes optimized for efficiency. Imagine a city where AI-controlled adaptive traffic lights dynamically manage congestion, and AV-only lanes ensure smooth transit, reducing accidents and maximizing efficiency. In Singapore, such initiatives are already being tested, offering a glimpse into an AI-coordinated urban future.

Rethinking Suburban and Rural Connectivity

As cities evolve, their impact extends beyond urban centers. In rural communities, AVs could bridge the gap between isolation and opportunity. Autonomous ride-sharing services will make personal vehicle ownership less necessary, turning once-distant villages into well-connected hubs of economic and social activity. Consider Raj, a retired teacher in a rural town, who can now visit the city independently, without relying on infrequent public transport, thanks to an on-demand AV service.

Sustainability and Equity

Parking structures of the future could integrate green roofs, absorbing urban heat and reducing carbon footprints. Charging stations powered by renewable energy will support net-zero emissions goals. More importantly, these transformations must prioritize equitable access, ensuring that AV-driven city redesigns benefit all communities, rather than exacerbating existing social divides.

The Technology Behind the Wheel

To fully appreciate the potential of autonomous vehicles, it's essential to envision the technologies that will drive them in the future. By 2030 and beyond, AVs will be powered by a new generation of AI and sensor systems, far surpassing today's capabilities. These advancements will not only make AVs safer and more efficient but also redefine our relationship with transportation.

Quantum Machine Learning

By 2030, quantum computing will revolutionize machine learning, enabling AVs to process data at unprecedented speeds. Quantum algorithms will allow AVs to analyze trillions of driving scenarios in real time, predicting risks and making decisions with near-perfect accuracy. This leap in computational power will eliminate latency, ensuring that AVs can navigate even the most chaotic environments, such as bustling city streets or unpredictable weather conditions, with ease.

For instance, a future AV powered by quantum machine learning could instantly process sensor data to anticipate a pedestrian stepping onto the street, adjust its trajectory, and communicate with nearby AVs to ensure a synchronized response, something impossible with today's computing limitations.

Advanced Sensor Systems

Future AVs will rely on next-generation sensors that integrate data from multiple sources:

- LiDAR 2.0: ultra-compact, low-cost LiDAR systems will provide 360-degree, high-resolution mapping of the environment, even in complete darkness or heavy rain.

- Biometric sensors: embedded sensors will monitor not only the vehicle's surroundings but also the health and emotional state of passengers, adjusting the driving experience in real time.

- Environmental sensors: AVs will detect and analyze air quality, temperature, and even seismic activity, providing valuable data for both passengers and city planners.

Imagine a scenario where a biometric sensor detects that a passenger is having a medical emergency. The AV could immediately reroute to the nearest hospital, alert medical personnel, and even adjust its driving style to ensure a smooth ride, demonstrating how future technology will extend beyond transportation into lifesaving applications.

Swarm Intelligence

By 2030, AVs will communicate with each other in real time using swarm intelligence, a decentralized system where vehicles share data and coordinate movements. This will eliminate traffic jams, reduce accidents, and optimize routes for efficiency. A compelling example of swarm intelligence in action is an AV highway system where vehicles move in coordinated "platoons." If a sudden obstacle appears on the road, AVs can react collectively, creating a dynamic buffer zone and rerouting traffic in milliseconds, ensuring smooth flow and enhanced safety.

Sustainability and Energy Harvesting

Future AVs will be powered by renewable energy sources, making them not only autonomous but also carbon-neutral, contributing to a greener planet. Innovations include:

- Solar-integrated AVs: vehicles equipped with solar panels charge batteries while in motion.

- Kinetic energy harvesting: systems convert the motion of the wheels into additional power, reducing reliance on external charging stations.

- Smart charging networks: AVs autonomously navigate to the most efficient charging stations, reducing wait times and optimizing grid usage.

For instance, cities like Oslo are already experimenting with roads embedded with wireless charging pads, allowing AVs to recharge while driving, eliminating downtime and enhancing energy efficiency.

Ethical AI in the Driver's Seat

Autonomous vehicles represent a quantum leap in transportation, but their true challenge lies in mastering morality. These intelligent machines must make split-second, life-altering decisions, balancing human safety, ethics, and cultural norms. How AVs navigate these dilemmas and earn public trust will define their role in shaping the future of mobility.

Learning from Human Behavior

AVs must continuously refine their decision-making by analyzing millions of real-world driving scenarios. In high-risk situations such as a pedestrian suddenly stepping into the road or an oncoming vehicle swerving unexpectedly, how will AVs respond?

For example, in dense urban environments like Mumbai, AVs must interpret unpredictable human movement, such as pedestrians weaving through traffic, whereas in Sweden, adherence to strict traffic discipline allows for more predictable decision-making. Real-world machine learning trials are already underway in multiple cities, training AVs to adapt to these diverse driving cultures.

The Trolley Problem in Motion

One of the most debated ethical dilemmas is the autonomous vehicle version of the "trolley problem": if an AV must choose between swerving into a wall (risking its passengers' lives) or colliding with pedestrians, how should it decide? While this debate has remained theoretical for years, recent research by MIT's Moral Machine Project has revealed how different cultures approach these ethical trade-offs. In Western societies, where individual rights are highly valued, AI tends to follow a utilitarian approach minimizing harm regardless of social roles. Meanwhile, in Eastern cultures, particularly in countries like China and Japan, ethical frameworks may incorporate Confucian principles, prioritizing the protection of elders or individuals with societal responsibilities. Imagine a real-world AV scenario in Germany where regulations mandate that AI must not differentiate between individuals based on age, status, or identity ensuring unbiased decision-making. By contrast, in a future scenario in Japan, an AV might be programmed to prioritize elders in an unavoidable accident due to cultural respect for hierarchy. These differences highlight the massive challenge of programming a globally accepted AI ethics framework for AVs.

Trusting a Machine to Make Life-or-Death Decisions

Despite AVs' ability to process vast datasets in milliseconds, public trust remains a significant barrier. How will passengers feel about riding in a vehicle that decides who lives and who dies?

A 2024 survey in San Francisco, where pilot AV programs have been tested extensively, revealed that 42% of participants felt uneasy about ceding complete control to an AI in high-risk situations, despite recognizing its potentially superior decision-making capabilities. This suggests that even if AVs statistically reduce accidents, people may still struggle to accept an emotionless machine making moral choices on their behalf.

A real-world case study from Waymo's 2023 testing phase in Phoenix, Arizona, demonstrated this challenge: an AV, when faced with an oncoming reckless driver, prioritized the safety of its passenger by making an evasive maneuver, leading to minor injuries for a cyclist on the sidewalk. The ensuing public debate questioned whether the AV made the "right" choice and how much transparency is needed in AI decision-making.

To address these concerns, manufacturers and policymakers must work together to build public trust, ensuring that AVs not only make logically sound decisions but also communicate their reasoning.

The "Empathy Engine"

Trust is fundamental to AV adoption. Future AVs must not only make ethical decisions but also communicate their intentions clearly.

Imagine a crosswalk where pedestrians hesitate, unsure if an AV will stop. How does the vehicle indicate its decision?

Upcoming technologies like interactive LED displays or AI-driven voice alerts could allow AVs to "signal" their thought process in real time, making their actions more predictable for humans. For example:

- A pedestrian approaching a crosswalk might see the AV project a visual cue onto the ground, indicating it is stopping.
- A vehicle could use an audible alert to communicate an emergency maneuver, helping nearby road users react accordingly.

As AVs become increasingly autonomous, their ability to foster transparency and build emotional connections will determine whether people truly trust them.

Cultural Adaptability

Driving behaviors vary drastically across the world, and a one-size-fits-all AI model will not work. Recent AV pilot programs:

- In Japan have shown a tendency toward cautious, measured driving styles, respecting pedestrian right-of-way always.
- In the US, where assertive driving is common, have seen AVs struggle to integrate due to overly conservative decision-making, leading to unexpected traffic slowdowns.
- In India, where fluid traffic rules and high pedestrian density make for a challenging environment, are testing AVs with hyper-adaptive AI that can interpret non-verbal cues from pedestrians and other drivers.

As AVs promise unparalleled safety, they also introduce new ethical, psychological, and cultural challenges. While AVs can reduce accidents dramatically, society must grapple with the reality of trusting machines to make moral decisions.

Moving forward, manufacturers, ethicists, and policymakers must work together to establish clear ethical guidelines. The biggest question isn't just how AVs will make decisions but whether humans are ready to accept those decisions.

Cybersecurity

As AVs integrate deeper into transportation systems, cybersecurity emerges as a critical pillar of trust. A connected vehicle is only as strong as its weakest link, and with AVs relying on real-time data exchanges, AI algorithms, and cloud-based systems, they become prime targets for cyber threats. Securing these digital highways is as essential as protecting physical roads.

The Threat Landscape

Imagine a scenario where a hacker infiltrates an AV fleet, rerouting vehicles, disabling essential safety mechanisms, or exploiting vulnerabilities in over-the-air (OTA) software updates. The consequences could be catastrophic, leading to traffic disruptions, data breaches, or even life-threatening accidents.

Recent research shows that automotive cyber attacks have increased by over 225% in the last five years, with vulnerabilities found in vehicle-to-everything (V2X) communication systems. As AVs become more interconnected, governments and manufacturers must work together to enforce cybersecurity standards before mass deployment.

AI-Driven Cyber Defense

The same AI that enables autonomous navigation can also serve as a cyber guardian. Machine learning algorithms continuously monitor network traffic, detect anomalies, and identify potential intrusions before they cause harm. Future AVs will integrate:

- Self-healing AI systems capable of isolating compromised components in real-time to prevent system-wide breaches
- Dynamic threat response mechanisms where AVs can reroute, shut down, or request human intervention when cyber risks are detected
- Collaborative cybersecurity networks, where AVs share threat intelligence with manufacturers, cybersecurity firms, and regulatory agencies to proactively mitigate risks

Blockchain for Secure Data Integrity and Transparency

AVs generate terabytes of data daily, from passenger behavior to route optimization and sensor readings. Ensuring data integrity is paramount, as compromised data could lead to dangerous decision-making errors. Blockchain technology provides tamper-proof records because of the following:

- Its vehicle logs, software updates, and AV decision-making processes are immutable and verifiable.
- Its multi-layered authentication prevents unauthorized access to AV data.
- With blockchain, governments and manufacturers can conduct transparent security audits, boosting public trust in AV decision-making.

The Future of Secure Autonomous Mobility

The road to AV adoption isn't just about smarter vehicles, it's about safer digital infrastructure. Achieving cyber resilience requires a unified effort between automakers, policymakers, cybersecurity firms, and AI researchers. By integrating AI-driven defense mechanisms, blockchain integrity, and a harmonized global cybersecurity framework, AVs can evolve from vulnerable targets to digital fortresses, ensuring trust, transparency, and safety in the autonomous era.

Policy and Regulation

The role of policymakers will be critical in shaping the future of AVs. Governments must establish standardized regulations for AV testing and deployment, ensuring safety and interoperability across borders. Additionally, liability frameworks must be updated to address questions of accountability in the event of accidents involving AVs. International collaboration will be key to creating a cohesive global framework that supports innovation while safeguarding public trust.

Automation and the Future of Work

Automation, over a long enough period, will replace every non-creative job. That means all our basic needs are taken care of, and what remains for us is to be creative, which is really what every human wants.

—Naval Ravikant

As automation and autonomous systems redefine industries, the workforce stands at a pivotal crossroads. While some fear job displacement, others envision a future of unprecedented creativity, efficiency, and opportunity. The impact of AVs extends far beyond the transportation sector, influencing manufacturing, logistics, retail, urban planning, and even healthcare. To adapt to this transformation, industries must embrace reskilling, create new job roles, and establish policies that ensure AI-driven automation serves human progress rather than replaces it.

Disrupting Traditional Roles

AVs are poised to disrupt millions of traditional jobs in industries like trucking, taxi services, and delivery logistics. With the rise of autonomous ride-hailing services and self-driving freight trucks, the demand for human drivers is expected to decline significantly.

However, this shift does not necessarily mean mass unemployment—rather, it signals the creation of new, tech-driven roles. For instance:

- Truck drivers may transition into fleet managers, remotely overseeing AV operations and ensuring efficient logistics coordination.

- Taxi drivers may evolve into mobility concierges, offering personalized travel assistance within autonomous transport ecosystems.

- Delivery personnel may shift toward last-mile logistics and AV maintenance, ensuring smooth operations in complex urban environments.

- Governments and companies must invest in large-scale retraining programs to reskill affected workers, enabling them to take on higher-value roles in AV ecosystems.

The Road Ahead

As the road to autonomy unfolds, the responsibility lies with us to shape its impact. Policymakers must advocate for ethical frameworks, innovators must push boundaries responsibly, and individuals must embrace the opportunities this transformation offers. The challenge isn't just about advancing technology; it's about ensuring that it uplifts humanity. In this defining moment, the choice is clear: will we harness AI to build a better world, or let its potential remain untapped? The future awaits our decision.

Engage with the Future

What does your ideal commute look like? How could an autonomous vehicle enhance your daily life? The road to autonomy is not just about technology; it's about reimagining how we live, work, and connect. The future is in our hands, let's shape it together.

About the Author

Pramod Patke is a visionary technology leader and product manager—strategy and tactics, with over 14 years of experience in autonomous systems architecture, product management, and intelligent design, shaping the future of mobility. He has collaborated with leading vehicle manufacturers worldwide, driving AI-powered transformation in the automotive industry.

As an expert system architect, Pramod has built high-performance architectures that enhance autonomy, predictive capabilities, and product scalability. His deep expertise in AI-driven innovation, data-driven decision-making, and machine learning, combined with his strategic mindset, allows him to bridge the gap between advanced engineering and business-driven solutions.

Passionate about the future of AI in transportation, he envisions a world where intelligent systems revolutionize mobility, making it safer, smarter, and more efficient. As an author, Pramod explores the transformative impact of AI on mobility, society, and human-machine collaboration, inspiring professionals and enthusiasts to navigate the AI-driven future with confidence and curiosity.

Email: pramodpatke@gmail.com

LinkedIn: https://www.linkedin.com/in/pramodpatke

CHAPTER 24

FROM RESISTANCE TO REVOLUTION: HOW ORGANIZATIONS CAN THRIVE WITH AI

By Carrie Purcell, MA, MBA
CEO, Insight Research Group & AI Labs
Toronto, Ontario, Canada

AI is not going to replace humans—but humans with AI are going to replace humans without AI.

—Dr. Karim Lakhani

Today, we have at our fingertips some of the most powerful AI tools in history, and they are easier to use than ever. While the foundations of AI have been evolving for decades, breakthroughs in computing power, cloud infrastructure, and data availability have propelled us into a new era. AI already plays a role in our daily lives, unlocking our phones with face recognition, filtering out spam emails, and suggesting what to watch next on Netflix. We now have unprecedented predictive power and efficiencies within reach. We have a chance to reimagine our roles, stepping into the director's chair of our own AI-powered teams.

If you're a business leader or AI enthusiast looking to enhance your organization with AI, this chapter is for you. We'll explore:

- Risks of Inaction
- Activating the AI Mindset
- Technology Adoption Theory
- Identifying Your AI Champions
- My 5-Step AI Blueprint for Seamless Integration

First, What Is AI and Why Does It Matter?

We'll define AI broadly here, borrowing from Melanie Mitchell's *Artificial Intelligence: A Guide for Thinking Humans*: "AI is the effort to automate intellectual tasks normally performed by humans," tasks such as problem-solving, pattern recognition, decision-making, and even creativity. This broad field encompasses the algorithms in traditional machine learning, the neural networks of deep learning, and the rapidly evolving field of generative AI.

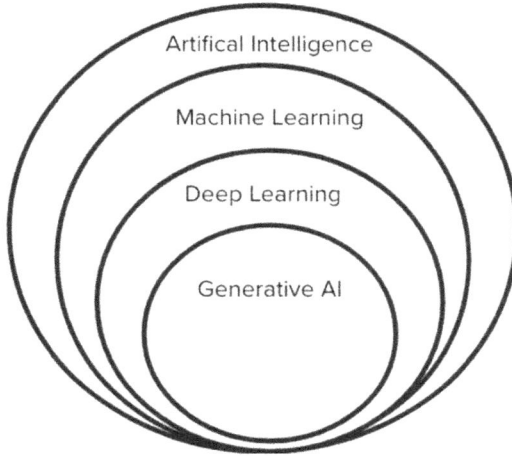

Unlike pre-programmed software, AI continuously refines its responses, evolving over time, and it is already transforming industries. In healthcare, AI is revolutionizing diagnostics and drug discovery. In finance, it's enhancing fraud detection and algorithmic trading. In manufacturing, AI-powered robots optimize production lines. Retailers use AI for personalized recommendations, and customer service chatbots are becoming increasingly sophisticated. So, what's next for your industry, your company? And importantly, what happens if you don't respond to the AI revolution?

Risks of Inaction: The Dystopian Storytelling Framework

To answer this, we can look to history for cautionary tales. Here's one for you:

> *Once upon a time, Jamie relied on taxis to get around the city, waving them down on busy streets or calling a dispatcher for a ride. And every day, Jamie paid metered fares, sometimes waiting longer than expected. Until one day, Jamie downloaded Uber and realized she could book a ride instantly, track the driver's arrival, and see the price upfront. And because of this, Jamie started using Uber for its*

convenience and affordability. And because of this, traditional taxi services saw fewer passengers like Jamie and struggled to compete with the new on-demand model. Until finally, taxis became a niche service while rideshare apps dominated urban transportation. And ever since that day, Jamie—and millions of others—have changed how they move around cities, choosing rideshares.

AI plays a key role in rideshares' ongoing market dominance over the taxi industry, enabling dynamic pricing, route optimization, and customer matching while the taxi's rigid pricing models and reliance on human dispatch cannot keep up.

But how can we anticipate these changes in our own businesses and adapt accordingly? The dystopian storytelling framework I'll share below (first introduced to me by the brilliant Ari Popper from SciFutures) allows us to safely imagine disruption (or destruction), through a visioning exercise which will very likely motivate change. Give it a try. Find 20 minutes today to follow this example:

Imagine a world in which your company has been left behind. You've missed critical paradigm shifts. Your customers have made different choices because you have failed to act on the potential of generative AI. Imagine the worst outcome.

You're writing about your character who is one of your company's key stakeholders introduced to radical disruption(s) that has put you out of business. Show the disruption and the consequences:

Once upon a time (in five to ten years) ...

And every day ...

Until one day ...

And because of this ...

And because of this ...

Until they never thought of ...

And ever since that day...

Writing this may be grim. My first try was downright depressing as I quietly considered the massive risks to the digital agency for which I was VP of digital and games—rapid, real-time, automated development would displace entire departments. Hyper-personalization would transform digital and gaming experiences on the fly, and it felt like it was all happening faster than we could test it, retrain, or rethink.

I wanted to run and hide, but I knew we had an opportunity here. I needed my team to see what I had seen. So, I assigned my Director team this same exercise. And they took my fears one level deeper, revealing exactly where our biggest vulnerabilities lay. Once I saw that and saw the future our competitors were developing, I could reduce the risk. It became clear how we could adapt and what I would need to change to transform our business into an AI-enhanced success story.

AI is going to be our secret weapon, offering the competitive advantage that the taxi industry failed to see, but that you will anticipate—and in this chapter, I will take you through the process to assess your workplace AI readiness and take the next steps. As the oft quoted Harvard professor Karim Lakhani says: "AI is not going to replace humans, but humans with AI are going to replace humans without AI." Now is the time to build a utopia where, yes, jobs will evolve and, yes, your business will change, but you will not be left behind. Now is the time to activate the AI mindset.

The AI Mindset

Organizations need to rethink what success looks like in an AI-driven world. The AI mindset is about strategic delegation—leveraging AI to handle repetitive and data-intensive tasks so that human effort can focus on creativity, leadership, and complex problem-solving. Pascal Bornet's *Irreplaceable: The Art of Standing Out in the Age of AI*, describes this shift saying:

> *Embrace delegation, especially to AI … it's about playing smarter. AI can … analyze mountains of data, and support or*

automate your decisions, freeing you to focus on what really needs your human touch ... it's about investing in gradually building a trusted collaboration with AI ... Letting AI take the wheel on some tasks doesn't diminish your skills, it amplifies them.

Amplify your human expertise. This gradual integration of AI allows organizations to build trust in automation, refine workflows, and eventually redesign entire business processes.

To do this successfully, change management expert Phil Lupton explains his approach and tips on a recent episode of *The AI Business Leadership Podcast*:

1. Develop AI literacy.
2. Build trust in AI systems.
3. Reduce fear of job insecurity using AI as an augmentation tool.
4. Encourage experimentation.
5. Find AI champions to lead adoption efforts.
6. Run pilot programs.
7. Demonstrate benefits, celebrating small wins.
8. Create cultural shifts, leading by example.

He states 10% to 15% of today's leaders are actively using AI while 50% to 60% of students and future leaders are engaging with AI tools. So, if you're new to AI, you're not alone—but you are already tackling step one, your AI literacy, by reading this.

You don't have to be the one designing AI-powered solutions. Empower AI champions and give them the room to experiment and refine AI initiatives. Let's look at Spotify as an example. Initially a music streaming service, Spotify competed primarily on its ability to let users create and save playlists. AI allowed Spotify to evolve into one of the most powerful predictive and personalized recommendation engines in the world. As described by Oskar Stål, Spotify's chief technology officer at the time:

When your predictions are accurate enough, something happens. You cross the threshold where you should actually rethink your whole business model and product based on machine learning ... with Discover Weekly ... We switched our vision from even better tools to "playlist yourself" to "you should never have to playlist again."

Spotify has an innovation ethos which embodies the AI mindset. It's about rethinking what success looks like utilizing AI capabilities without being restricted by current tools, budgets, trust concerns, or data privacy roadblocks (we'll get to all of those later in the systematized AI blueprint).

First, let's develop the vision. How can you get started in your organization? Here are some great resources for Lupton's first four tips:

- Newsletters such as "The Neuron" or Medium's AI-focused publications

- Podcasts, books, and articles such as *Eye on AI*, The Economist's AI coverage, Harvard Business Review's AI research, books like *AI Superpowers* by Kai-Fu Lee and *Prediction Machines* by Ajay Agrawal

- Tool testing such as ChatGPT for content generation, Gamma for designing presentations, Runway ML for video editing, Perplexity AI for research and information retrieval

Encouraging AI literacy and experimentation means creating time for it. Podcasts are great during a commute or over lunch. Newsletters may be the first email you and your teams read each morning. Tool testing like ChatGPT and Gamma will create efficiency gains, which can be redistributed to additional tool testing, article reviews, books, online events, conferences, etc. Jump in and encourage your teams to—it will save you time in the long run.

AI Adoption Theory and Champions

Now to find your AI champions. It is helpful here if we dig into technology adoption theory—the focus on my doctoral research. Everett Rogers' Diffusion of Innovations theory provides a framework for understanding how innovations spread within societies (later expanded by Geoffrey Moore for modern businesses) through five stages:

1. Innovators
2. Early adopters
3. Early majority
4. Late majority
5. Laggards

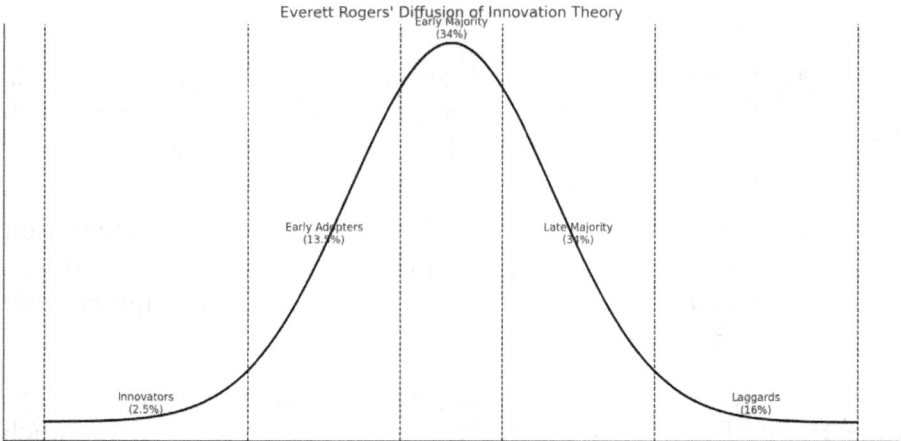

Everett Rogers' Diffusion of Innovation Theory

Adoption for disruptive technologies progresses through each of these stages adding more and more users until the technology becomes mainstream. I like to think of these stages as personas whom you can identify with and within your organization:

(1.) The "innovator" is embodied by Alex, the *Visionary*, a CTO at a mid-sized tech firm who frequently experiments with AI-powered analytics and automation, looking for the next major leap to give the company a competitive edge.

- Traits: future-focused, risk-tolerant, constantly experimenting with AI tools and models
- Role in AI adoption: identifies and experiments with cutting-edge AI applications, setting the stage for broader adoption

(3.) My "early adopter" is Jamie, the *Change Agent*—yes, also the Jamie from our taxi-Uber story earlier in this chapter—who is a marketing manager and uses AI-driven insights to refine customer segmentation and personalize outreach. Jamie loves testing new AI-powered tools and sharing success stories.

- Traits: enthusiastic about AI, influential in their teams, persuasive, and quick to adopt new workflows
- Role in AI adoption: advocates for AI's benefits, influencing colleagues and ensuring early successes

(3.) The "early majority" is represented by Taylor, the *Pragmatist*, an operations manager who is open to AI but needs clear ROI before approving any new technology. They rely on pilot programs and user testimonials before recommending broader adoption.

- Traits: cautious optimist, data-driven, seeks evidence before committing to AI adoption
- Role in AI adoption: provides critical mass and scalability once convinced AI has proven value

(4.) Our "late majority" is Morgan, the *Traditionalist*, and our finance director who is hesitant about AI-driven forecasting tools but slowly comes onboard after seeing industry-wide adoption and internal success stories.

- Traits: risk-averse, skeptical of change, prefers proven systems, and requires substantial reassurance

- Role in AI adoption: willing to adopt AI once it becomes the norm but requires structured onboarding and clear ROI

(5.) And our "laggard" is Chris, the *Skeptic*, a veteran engineer who prefers traditional methods over AI-driven automation. They question AI's reliability but will eventually adapt if required by company policy.

- Traits: resistant to change, tech-averse, often critical of AI's role in the workplace
- Role in AI adoption: adopts AI only when mandated or when legacy systems become obsolete

Which one are you? As you start to recognize these personas within your organization, you can tailor your AI adoption strategies to engage each group effectively.

I used to be Morgan, the *Traditionalist* and "late majority," and proudly so. I waited and watched, a risk-averse skeptic, as everyone I knew joined Facebook. Was it going to stick around, have proven value, could I trust it? When I eventually joined, I was hesitant to share photos or identify I was out of country or show legal information on degrees. And my instincts weren't wrong regarding privacy, security, and the ethical use of data, something we address carefully in step 1 of the AI Blueprint.

When I got my first Apple Watch, I was promoted to Taylor, *The Pragmatist* and "early majority." I was a Fitbit user who heard Apple Watch skeptics report that while it could do everything my Fitbit could, plus make phone calls, send texts, and play my music, it had no clear use case and, therefore, wouldn't survive. I waited a few Apple Watch generations seeing no clear use case, but by series 5 I decided to just try it out. The heart rate monitor was superior to my Fitbit, I could play music and easily set timers, decline phone calls, and respond to text messages using voice control. My entire Apple ecosystem was synced between my computer, phone, and watch, and I loved it!

Soon after, I jumped squarely into the "innovator" persona. I became Alex, the *Visionary* and *Innovator*, when I co-founded and launched my metaverse for education platform in 2019. I was interested in creating more engaging and immersive online learning environments using video game technology, but we were way ahead of what the market was ready for—until April 2020 when everything changed. You'll remember this period. As people dialed into school and work through COVID-19 isolations, our innovative product found product-market-fit and experienced massive growth. Digital transformation was occurring rapidly around us, and my company was in the right place at the right time. I observed the technology adoption cycles accelerate with an unprecedented trigger, a global pandemic, and I knew it was an area I needed to understand further, leading to my research today.

Whoever you are today, you are surrounded by all of these personas. And you can move through the stages as I did. Find your Alex and Jamie to help you lead the change. Let them experiment. This will drive engagement, trust in AI, and faster adoption. Taylor will jump on board next. Morgan and Chris will keep things in check, but this tension will be healthy as your organization navigates increasingly larger transformations.

This bottom-up approach encouraging innovators within your organization to experiment is the ethos of how Spotify innovates and is an essential part of AI adoption. Top-down AI initiatives, on the other hand, ensure strategic alignment, regulatory compliance, and enterprise-wide deployment. These are used for mission-critical systems like SAP or AWS and large-scale AI implementations in highly regulated industries like finance or healthcare. AI adoption in organizations succeeds when bottom-up innovation and top-down strategy work in tandem, so let's look more closely at top-down transformation.

The AI Blueprint

You are ready. You have the vision from your dystopian storytelling exercise. You've worked on your AI literacy and identified your AI champions. You have the AI mindset and want an AI-first approach to your business. Let's start by identifying the right opportunities. At Insight Research Group & AI Labs, we use the AI blueprint framework to uncover the right AI opportunities for your organization. I'll share the five-step approach, so you can start today.

Step 1: Identify the Right AI Opportunity

- Define the business problem(s) AI can solve.
- Assess the problem category and most appropriate AI for the job.
- Look for high-impact, high-feasibility, low-risk use cases.

In a recent large-scale healthcare project, my team conducted stakeholder interviews across seven groups, identifying over 100 potential AI use cases. Using a Pain and Gain matrix, we ranked opportunities based on revenue potential, positioning, and sustainability while weighing ethical, legal, budgetary, and technical challenges rigorously. This data-informed approach helps leaders prioritize AI initiatives, classify problems by autonomy level, and assess business impact before choosing the right AI opportunities.

Step 2: Align AI to Business Strategy

- Ensure AI initiatives align with company goals (e.g., efficiency, customer experience, revenue growth).
- Secure leadership buy-in to support AI adoption.
- Assign an AI sponsor to champion integration.

We identified that our healthcare client's best opportunity was a predictive analytics system that would replace an antiquated model and connect more stakeholders in the final solution. The anticipated 20% efficiency gain is an 11-figure savings in this case—we want a clear ROI in step 2. Also think about your personas and who will help move this forward or cause resistance.

Step 3: Assess Data Readiness and Infrastructure

- Evaluate data availability, quality, and accessibility.
- Identify gaps in AI talent, tools, and infrastructure.
- Ensure compliance with privacy and regulatory requirements.

Here we assess available data sources and types (structured, unstructured, hybrid), creating a large matrix and developing the data collection plan in three stages: what's (1) available, (2) easy to collect, and (3) hard to collect with projected timelines. We build our MVP with (1) available and (2) easy to collect data updating the model in the next phase with (3) hard to collect data. We also consider data cleaning and bias reduction as well as a scalable data management strategy with cloud storage, robust pipelines, and governance frameworks for security and compliance. With a clear data foundation, we also define the team requirements in this step.

Step 4: Develop and Validate the AI Solution

- Choose between off-the-shelf AI tools vs. custom AI development.
- Run pilot projects to test AI performance.
- Gather user feedback and refine the model.

The core of this solution would be an LLM trained on anonymized medical records, handling our unstructured data well and enabling us to accelerate MVP development. We evaluated open-source LLM models for modification and testing, identifying an option with higher diagnostic accuracy than human physicians. We set clear performance metrics reaching 95% accuracy for a decision support tool.

Step 5: Deploy, Monitor, and Scale AI

- Implement AI in production workflows.
- Establish AI performance metrics and monitoring systems.
- Train employees and drive cultural adoption.
- Scale AI solutions across teams and functions.

Now we onboard pilot groups executing our first phase of deployment with continual monitoring and retraining at predetermined intervals. Based on our results, we will progressively execute on subsequent development phases (continual data collection, model training, and scaling).

Conclusion

Deployment gets us to launch, which any good Product Manager will tell you is just the beginning. For your business, it is the beginning of a whole new world, a new way of operating, re-exploring your responsibilities, capabilities, and teams.

For your product, it's V1 and will be iterated, developing future features for continual improvement. But you know AI adoption is not just about the tools—it's about rethinking your organization using the AI mindset and the AI blueprint to transform vulnerabilities into a utopian vision for your company's future. With this, you will ensure your organization is positioned to thrive in the AI revolution.

About the Author

Carrie Purcell is an AI scholar and emerging technology strategist dedicated to transforming businesses through innovation. A sought-after consultant, speaker, and professor, she bridges the gap between cutting-edge AI research and real-world applications. Recognized as one of Toronto's top female business leaders and an award-winning entrepreneur, she has been honoured for her visionary contributions to technology and leadership.

As a doctoral candidate specializing in Generative AI, Carrie explores the intersection of artificial intelligence, business strategy, and digital transformation in her AI adoption and responsible AI research. She co-founded Adatpika, a VR gaming company, and today leads Insight Research Group & AI Labs, helping organizations harness AI for competitive advantage.

With a master's degree from the University of Toronto and an Executive MBA in Blockchain & Entrepreneurship, Carrie combines academic excellence with hands-on expertise. Carrie also has a three-part book series coming soon on machine learning, deep learning, and generative AI.

Email: cpurcell@insightresearch.tech

Website: www.Insightresearch.tech

CHAPTER 25

FROM FICTION TO FUNCTION: HARNESSING AI TO SOLVE REAL-WORLD CHALLENGES

By Philipp Ramjoué
Consultant & Trainer, Edge AI & IoT
Arusha, Tanzania

> *Be the change you wish to see in this world.*
> —Mahatma Gandhi

From Science Fiction to Reality—What AI Must Be Able to Do Today

Artificial intelligence has long fascinated people as the subject of different movies. In admiring amazement we saw the self-driving car "K.I.T.T." from *Knight Rider* whizzing across the screen autonomously in 1982. In the 2004 film *I, Robot* we sympathised with the humanoid robot Sonny, which made us aware of the moral aspects of AI. Today such systems are already reality or close to completion. We talk to intelligent voice assistants, get help with everyday routines like booking a table at a restaurant. Maybe we'll be picked up by a robot taxi from Waymo or just use our own Tesla with autopilots that are only slightly inferior to "K.I.T.T.". The first testers are already served at home by humanoid robots from Tesla and bring the idea of Sonny to life.

But which problems are really solved by using these systems? Have we not been able to overcome these challenges so far based purely on human resources, or are we developing systems to end in themselves "just because we can"? Where do these systems come from, what is the personal responsibility of all of us, and how do we use AI tools to solve the real problems of our world in a sustainable way?

Harnessing AI to Solve Real-World Challenges

The range of modern AI systems and services is growing rapidly. New achievements are constantly presented and more complex "problems" are solved without limits to the possibilities. But at this point it is worth taking a deeper look and asking: which problems are we currently addressing and what should we actually be focusing on?

AI from the Right Pocket to the Left

If we start looking where these systems come from and for what purpose they are used, even superficial research reveals that almost all solutions are basically produced and sold by private companies with the pure "intention of making a profit", explicitly by the big players in Silicon Valley. These companies are little known for their humanistic and sustainable approaches, but more for being at the forefront of technology and regularly redefining the limits of what is possible. The motto is "higher, faster, further" and not "sensible, conscious, and sustainable for the good of this world". However, alongside our admiration for what is technically feasible today, let's ask ourselves:

which challenges were unsolvable in the past and are solved now thanks to these systems?

The End of the Daily Drama of Switching on the Light

With all the solutions currently on the AI market, it is critical that people question the meaningfulness and true benefits of technical systems, especially in the case of resource-intensive solutions like AI. More manufacturers are developing intelligent assistants and integrating them into our everyday lives with the promise of increased convenience, personalised services, and greater security. The term "smart home" includes intelligent lights, self-thinking refrigerators, as well as intelligent security systems that at least go beyond replacing small household tasks, but a closer look still raises the painful question: what problems are we actually solving here? Have we been overwhelmed by switching on the lights ourselves in the past or by drawing up a shopping list?

And immediately we hear the voice of the proponents: "These technologies simplify our everyday lives". Let's agree with them and accept that for many people, saving time by not switching on lights manually makes a real difference. But what might be lost as a result? The WHO is already warning that people are moving too little. Walking to the light switch will certainly not solve the problem, but it is symptomatic of our times. More comfort leads to inertia and demonstrably to problems such as obesity or cardiovascular problems.

And once again, the advocates are loud and point out, "We can use the time we gain much more sensibly", for example, while travelling in an autonomous car or when robots with AI on board finally take over physically hard or monotonous work. At the same time, there is a feeling that people's daily working hours are going up instead of down, burn-out rates are increasing, serious mental illnesses are on the rise, and people are looking for meaning in their lives like never before.

And the "yes, but" game continues: "Autonomous cars are also so much safer!" Do we really have a safety problem on the road today that urgently needs to be solved, or should we and AI be much more concerned with solving the excessive amount of private transport, CO2 emissions, the lack of connections between rural regions and local transport, and the awareness of the need for transforming the individual mobility sector? The game could go on forever with the same result.

These statements can now seem like an absolute plea against any use of AI from a developer of AI (me) in a book about AI (this book). What this change of perspective is intended to encourage is an awareness of the challenges associated with the use of AI, the ability to look at the flipside and to question: do we really need it? And, of course, we can then also decide in favour of using the above-mentioned solutions, but with awareness of all facets and by adding the question of meaningful use beyond this, so that in the future we not only push the boundaries of what is technically feasible, but also we aim to overcome problems that we as humanity have not been able to solve on our own yet. It is, therefore, worth taking a look at the many real challenges of our time.

AI for Real—Solving the True Challenges

The "Global Risk Report 2025" identifies the following as the first four risks of the next ten years: extreme weather events, biodiversity loss and ecosystem collapse, critical change to Earth systems, and natural resource shortages. All of these points are "environment" risks. This is not about our living standards deteriorating or life on Earth becoming more uncomfortable. We are simply talking about the fact that sooner or later humanity's lights will be switched off, without any smart home application. These risks are the real challenges of our time that need to be addressed. So let's discover what AI, with its unique analytical and predictive skills, is really capable of and how it can sustainably improve, if not save, our lives.

AI to Counter Extreme Weather Events

Many experts agree that the climate has changed and will continue to do so. People see their livelihoods threatened by this or even have to leave their homes because areas are no longer habitable or agriculture is no longer possible. AI could make a difference in advanced weather analyses and forecasts. Models can process large amounts of data simultaneously and provide earlier warnings of extreme events. Using satellite and sensor data, AI recognises early indicators for hurricanes, floods, and forest fires.

Google's "Flood Forecasting Initiative" predicts flooding in endangered regions up to seven days in advance. AI has the ability to not only predict natural disasters, but also to optimise targeted evacuations and emergency measures, crucial for people in these areas. These systems protect humans in the event of disasters, but climate change also affects people and animals on a daily basis. Unpredictable rain and heat waves lead to crop failures and water shortages. AI farming apps can help with real-time local weather, soil and moisture analyses that help with targeted and needs-based irrigation of plants, rounded off by automated growth controls and planting recommendations, enabling sustainable agriculture. But even if we manage to adapt, humans are ultimately not alone, but part of the ecosystem and, therefore, dependent on it.

AI to Reduce Biodiversity Loss

The loss of biodiversity and increasing environmental pollution is a time bomb. It is often small changes that cause systems to tip over— and it's almost impossible for humans to recognise those connections. As already mentioned, processing extremely large volumes of different data and precisely finding correlations is a "core competence" of modern AI solutions. Questions such as "Which factors are accelerating biodiversity loss the most?" or "Where is the greatest lever to stabilise the system?" could be investigated with AI. We could use "directly in ecosystems" camera traps, drones, and IoT sensors with onboard AI (EdgeAI) to monitor endangered areas in real time, identify wild animals, detect illegal activities, such as poaching or deforestation, and alert rangers immediately. Rainforest Connection uses AI to recognise chainsaws in rainforests with audio recorders and initiate protective measures. Other systems predict where poachers will strike next. Protected areas are ecosystems of unimaginable size that are impossible for humans to monitor comprehensively.

Until now, we have only been on the lookout for environmental destruction and poaching, but now we have the tools to actively counter these threats in real time. Human-wildlife conflicts could be prevented, for example, in areas of tension between elephants and humans competing for habitat, where elephants cross settlements in East Africa on their migration routes, which have been the same for centuries, seeking out sweet maize fields and water reservoirs. AI can be used to warn people at an early stage or, even better, to prevent animals from entering farms at an early stage by using drones and intelligent wildlife cameras to analyse animal behaviour and predict routes. These technologies have a preventative effect and are one piece in the big puzzle of solving the loss of biodiversity, which is being exacerbated by advancing climate change.

AI to Stabilise the Climate

The Earth is a highly complex system, an interplay of atmosphere, oceans, soils, forests, and glaciers to enable life. Everything influences each other such that the melting of the polar ice caps, the acidification of the oceans, or the disappearance of the rainforests have global consequences. By analysing huge amounts of data from satellite images, sensors, and weather models AI can make more precise predictions—better than ever before. Traditional climate models only work to a limited extent, so the graphics processor manufacturer NVIDIA is developing the digital Earth twin "Earth-2", which can be used to precisely simulate how climate changes will affect us in the coming years.

These models enable taking targeted local and global measures, and providing that information to politicians and decision-makers. One of the biggest problems of our time is rising CO_2 emissions. Savings in industry and transport can have a great deal of leverage, but should be implemented in a sustainable and forward-looking manner. At the same time, the digital twin enables real-time monitoring and evaluation of the success of the measures. So instead of just slowing down climate change, we have the opportunity to stabilise the Earth system in the long term. However, when using all these systems, the huge consumption of resources should be focused on.

AI to Save Natural Resources

Modern AI systems are currently developed by large tech companies. This development is made possible by the use of gigantic server farms and data centres. Energy consumption on the scale of small towns is usual and ultimately leads to the emission of greenhouse gases. Maybe we are not able to avoid the high consumption of resources for development at present, but we can use AI to cover our electricity requirements in a CO_2-neutral way. AI-controlled smart grids enable the optimal distribution of wind and solar energy and control the intelligent charging and discharging of storage systems. AI can additionally analyse where and when particular electricity peaks occur and "serve" them. This allows the grid to be better utilised and losses to be reduced as Google DeepMind reduced the energy consumption of data centres by 40% by introducing AI-controlled cooling processes.

Additionally, the waste and contamination of drinking water should also not be neglected. AI can be used to analyse and monitor processes and make them more efficient as intelligent leakage detection algorithms in water pipes prevent the waste of valuable drinking water in cities and soil analyses in fields enable plants to be watered precisely.

With all the chances of using AI to solve real challenges we should think about resources AI itself uses to do "great stuff". One very promising innovation to reduce the usage of resources is EdgeAI/TinyML. Trained models can be "extracted" from the cloud and find a "home" on microcontrollers and mobile devices. With this achievement high-performance models can be operated on low-cost, highly available systems that are mobile and often energy self-sufficient, and can make predictions. The data does not have to be sent to the internet to be analysed. In combination with IoT technologies, a powerful tool is created. By processing on the device, a comprehensive, available broadband internet communication becomes obsolete, and other forms such as satellites or "low-power wide-area networks" can be used. This makes it possible to use AI in remote areas or protected ecosystems in a way that conserves resources. But who is using these solutions in the end? It's all about the people.

AI for/from the People

AI is a tool and, therefore, neither good nor bad, it's about how we use it. The real world is us. Finally, it's always people who make the difference. It's about awareness of what AI is and what it isn't, what AI can and can't do, what we get from AI and what we might lose, how many resources we invest and what we achieve as a result. People should be empowered and trained to answer these questions for themselves individually, starting with the youngest members of our society in schools. Through this AI can be used precisely where it makes sense and is expedient, where it is worth the investment of resources and where we, as humanity, benefit from it. From being an end in itself towards being a tool.

In the overall human context, there will always be people who have barriers for a variety of reasons and who will be denied access to everyday life or the labour market. So we have to transform from the boundless pursuit of profit and silo thinking towards a diverse society in which everyone can live well and take responsibility. The use of language AIs can reduce talking barriers, and previously unskilled people can be supported with augmented reality glasses to learn new jobs or work in unfamiliar jobs. Intelligent assistance systems can provide individualised support for people with disabilities or the elderly and enable them to lead self-determined lives. As a co-worker or co-pilot, AI can empower and enable people and accelerate successful integration and inclusion. In the end, we are one human race, and the challenges of our time do not stop at our own front door.

Conclusion: AI as a Pioneer for Real Solutions

After all these considerations, I would now like to let a modern AI have its say:

> "ChatGPT (4o), enclosed you will find my thoughts on the sensible use of AI for real challenges of our time. I would like to ask you, as a highly developed AI, to write the conclusion for this chapter".

> ChatGPT: "Artificial intelligence has long been a reality. But instead of just using it for convenience and profit, we have the opportunity to use it to tackle the biggest challenges of our time".

The examples in this chapter show that AI *can be more than just a luxury product*—it can protect ecosystems, combat climate change, utilise resources more efficiently, and improve the way we live together. It helps us to make informed decisions by recognising correlations that are too complex for humans. But AI is not an end in itself—its development follows the interests of those who create it.

The decisive factor is, therefore, *how* and *where* we use it. It should be used specifically for sustainable solutions instead of just maximising efficiency gains and convenience. Instead of replacing human capabilities, it can serve as a tool for solving problems—for example, in environmental protection and resource conservation. In doing so, we must also critically scrutinise its own energy consumption to ensure that it itself remains sustainable.

AI can be a real game-changer, but only if we use it consciously and purposefully in areas where we, as humanity, cannot make progress on our own. It is up to us to use it not just as a vision, but as a lever for real, sustainable change." With this in mind:

About the Author

Philipp Ramjoué is an AI and IoT consultant, specialising in EdgeAI, TinyML, and AI-driven solutions for real-world challenges. With a background in mechatronics and deep learning, he has worked extensively on AI, computer vision, and IoT applications, always with a focus on sustainability and practical benefits.

Philipp brings years of industry experience from Germany, where he developed AI-driven automation solutions, industrial IoT applications, and real-time embedded systems. His work included TinyAI for resource-efficient AI processing on hardware, as well as cloud-based AI solutions for sensor analysis and automation.

As a co-founder of Raingers GmbH, Philipp applies AI to nature and wildlife conservation, developing solutions such as AI-powered drones with thermal imaging for real-time wildlife monitoring. His goal is to use AI not just for technological progress, but to tackle pressing environmental and societal challenges.

Beyond development, Philipp is passionate about AI education and consulting, helping businesses and individuals understand and implement targeted, resource-efficient AI solutions. He has trained future AI managers, advised companies on AI-driven automation, and worked on TinyAI applications for low-power, on-device AI processing.

Philipp currently lives in Arusha, Tanzania, where he integrates AI with sustainability, focusing on mindful technology use and conservation efforts. His work is based on the principle that AI should not be an end in itself—it should be a tool for meaningful, real-world impact.

Email: philipp@ramjoue.ai

Website: www.ramjoue.ai

LinkedIn: https://www.linkedin.com/in/philippramjoue/

CHAPTER 26

THREE CRITICAL SKILLS TEAMS NEED IN THE AI ERA

By Daria Rudnik
Team Architect & Leadership Coach, Aidra.AI Founder
Tel Aviv, Israel

The ability to envision and drive change is just as important as the ability to work with technology. If you don't have both, you can't succeed in this world.

—George Westerman, Senior lecturer, MIT Sloan

How AI Is Changing the Way Teams Work

Whether we're ready for it or not, AI is transforming the way we work. I see it in every organization I work with—AI is reshaping industries, workplaces, and team dynamics at an unprecedented pace. And what does that mean for your team? How do you prepare them for a world where artificial intelligence plays an increasingly central role in decision-making and daily operations?

If you're reading this book, chances are you're already using AI to make your work more efficient. Many of my clients have started integrating AI tools into their routines, and I've even developed an AI-powered coach—Airda.AI—that supports them between sessions. However, optimizing personal productivity with AI is one thing. Managing AI's role in a team is another challenge altogether.

The real question is: how does AI change teamwork? AI is changing how people interact, make decisions, and work. It's not just about doing tasks faster or handling data better. While AI can help us be more efficient, it also brings new risks that might not be obvious right away.

Some teams thrive with AI, using it to enhance creativity and collaboration. Others, however, struggle. Curious to know the difference? Surprisingly, it's not about having access to AI. It's about knowing how to use it without losing essential human skills—things like critical thinking, intuition, and deep engagement.

That's what this chapter is about. AI isn't going anywhere, but how we integrate it into our teams will determine whether it enhances our abilities or weakens them. To make the most of AI while avoiding its pitfalls, we need to understand one of the most significant challenges it introduces: cognitive offloading in an AI-driven world.

Cognitive Offloading: What Happens When We Let AI Think for Us?

AI makes life easier. It handles repetitive tasks, organizes information, and even suggests decisions. But this convenience has a hidden cost—one that many teams don't recognize until it starts affecting their performance.

Cognitive offloading is relying on external tools to store or process information instead of using our memory and reasoning. We do this all the time in daily life. Instead of memorizing phone numbers, we save them in our contacts. Rather than remembering a grocery list, we jot it down in an app. Navigation apps mean we no longer need to map out a route in our heads before driving somewhere new.

These tools free up mental space, allowing us to focus on other things. And in many ways, that's a good thing. Offloading routine tasks can make us more efficient. But when it comes to deeper thinking, creativity, and decision-making, over-relying on AI can weaken our ability to process, retain, and meaningfully engage with information.

Teams that rely too heavily on AI can start to lose touch with their core work. The more people delegate thinking to AI—whether for customer insights, strategy, or decision-making—the less they actively engage with the work itself. Employees who previously mastered their domain may feel removed from it. They trust AI's output without questioning it. Over time, their ability to think critically and solve problems weakens.

The challenge isn't whether to use AI—it's about knowing when to let AI assist and when to stay mentally engaged. The teams that thrive in the AI era recognize this balance.

What Research Tells Us About Cognitive Offloading

The idea that technology changes the way we think is backed by science. Researchers studying cognition tell us that the more we rely on external tools to store and process information, the less we engage our memory and reasoning skills. One well-documented example is the Google Effect or digital amnesia. Studies have shown that when people know they can look up information later, they are less likely to remember it.[1] Instead of absorbing knowledge, they subconsciously assume they'll always have access to it. In other words, we trade deep learning for convenience.

This doesn't just apply to memorizing facts. Research has found that when people use AI or other digital tools to assist decision-making, they often overestimate their understanding of a topic.[2] It's easy to mistake access to knowledge for actual expertise. Just because an AI tool generates an insightful report doesn't mean the person using it fully understands the nuances of the data.

Even more concerning, studies suggest that when we consistently rely on technology for thinking, we weaken our ability to engage in deep problem-solving and critical analysis. This is especially relevant in workplaces where AI-generated reports, summaries, and recommendations have become routine.

When AI Makes Teams Less Engaged

Ana, the head of customer success at a leading cloud company, noticed a troubling trend—her customer success managers were showing signs of disengagement. This was puzzling because there were no obvious triggers. The company had successfully navigated a round of layoffs a couple of years prior while maintaining high morale, and business had been steadily growing. There were no major conflicts or disruptions that might explain this shift in engagement.

Particularly concerning was that managers who once knew their client accounts inside and out were struggling to recall basic account status details during meetings. While all the records were meticulously maintained, the deep, intuitive understanding of their accounts seemed to be fading. When we investigated the root cause, we discovered it lay in how their work routines had evolved after implementing an AI copilot. Managers now used AI to record meetings, generate reports, and create customer insights—all of which kept their records impeccable. There was a cost to that— managers were no longer mentally processing and internalizing the information. Their role had essentially been reduced to two functions: having conversations with clients and operating AI tools. They had fallen into the trap of cognitive offloading, where the tools meant to enhance their work inadvertently diminished their engagement and understanding.

Don't get me wrong. In no way am I suggesting that teams should not use AI to optimize their workflow. The point is that teams need to be mindful of what tasks they delegate to AI, when they collaborate, and when they use their human power to arrive at the best solution.

To help my clients like Ana, I've created a simple checklist for teams to analyze their workflow and tasks to determine which could be delegated to AI, where there are opportunities for human-AI collaboration, and what tasks are purely human responsibilities that should never be delegated or shared with AI.

AI Task Navigator: When to Delegate, Collaborate, or Lead

The goal of this AI Task Navigator is to ensure that AI enhances human work without replacing critical thinking and expertise. Use this simple step-by-step checklist to decide whether to:

- Delegate to AI (AI handles the task independently).
- Collaborate with AI (AI provides support, but humans refine and finalize).
- Keep it human-only (AI should not be involved).

This approach allows teams to stay engaged while still benefiting from AI's efficiency. Here's how it works:

Step 1: Answer These Yes/No Questions

1. Is this task part of your team's core function?

   - YES → AI-human collaboration (AI can assist, but humans lead)
   - NO → Move to the next question.

2. Does this task directly create value for stakeholders?

   - YES → AI-human collaboration (AI supports, but humans ensure quality)
   - NO → Move to the next question.

3. Does this task require emotional intelligence, ethics, or deep human judgment?

- YES → Human-only task (AI should not be used)
- NO → Move to the next question.

4. Is this a repetitive, rule-based, or data-heavy task?

   - YES → AI delegation (AI can fully automate this)
   - NO → Move to the next question.

5. Does this task require strategic decision-making or high-level problem-solving?

   - YES → AI-human collaboration (AI provides insights, but humans decide)
   - NO → Move to the next question.

6. Can AI generate useful insights or ideas for this task?

   - YES → AI-human collaboration (AI assists, but humans refine)
   - NO → Human-only task

7. Will AI enhance creativity (e.g., brainstorming, idea generation)?

   - YES → AI-human collaboration (AI helps, but humans finalize)
   - NO → Move to the next question.

8. Does your organization have the right data and tools for AI to perform this task reliably?

- YES → AI delegation (AI can take over this task)
- NO → AI-human collaboration (AI can assist, but humans oversee)

Step 2: Interpret Your Answers

- Mostly AI delegation → AI can fully handle this task.

  Example: Data entry, automated reports, simple customer inquiries

- Mostly AI-human collaboration → AI can assist, but humans should stay involved.

  Example: Strategy planning, product innovation, customer relationship management

- Mostly human-only → AI should not be used for this task.

  Example: Leadership decisions, conflict resolution, ethical considerations.

Final Check Before Delegating to AI

Before assigning a task to AI, ask:

- Have we reviewed AI-generated insights for bias or errors?
- Will AI improve efficiency without reducing quality?
- Are there clear processes for human oversight?

I do realize that AI is evolving fast, and some of these questions might not be as relevant in a few months—or even weeks. But I still want to offer them here because this checklist isn't just about answering specific questions—it's about fostering a thoughtful, intentional approach to AI. The real goal is to ensure that teams develop the skills they'll need to stay effective as AI becomes more integrated into their work.

With that in mind, let's focus on the three key skills that will set teams apart in the AI era—the uniquely human strengths that AI can't replicate and that will be essential for staying competitive in a rapidly changing world.

Key Skill #1: Cultivating an Adaptive Learning Mindset

Learning to adapt isn't just a nice-to-have anymore—it's crucial as AI keeps changing the game. The teams that thrive are the ones that embrace continuous learning, staying curious and open to change rather than feeling overwhelmed by it. A report by the British Council ranks learning agility as one of the most important skills for the AI era.[3] But this isn't just about taking an occasional training course or attending a workshop. It's about developing a mindset that actively seeks knowledge, experiments with new tools, and applies feedback to grow.

How to Build an Adaptive Learning Mindset in Your Team

1. Encourage hands-on learning.

Reading about AI is useful, but actively using AI tools makes learning stick. Create opportunities for team members to experiment with AI in a safe environment.

2. Integrate microlearning into daily workflows.

Instead of overwhelming employees with lengthy training sessions, introduce small, practical learning moments throughout the workweek. Short tutorials, peer discussions, or quick AI-assisted projects can be far more effective.

3. Use peer learning.

One of the most effective ways to learn is from each other. Pair team members with different levels of AI expertise, so they can share insights and best practices.

An adaptive team isn't afraid of AI. They see it as a tool to enhance their work rather than something to resist. But adaptability alone isn't enough. The next essential skill is collaborating effectively with AI rather than simply accepting its outputs at face value.

Key Skill #2: Redefining Collaboration in the AI Era

AI is great at handling repetitive tasks, processing vast amounts of data, and generating insights. But it can't replace human judgment, creativity, or the ability to interpret information in a nuanced way. That's why teams need to rethink collaboration in an AI-driven workplace. It's no longer just about working well with other people—it's about learning how to work effectively with AI. The key question is: how do we ensure AI enhances human collaboration instead of replacing it?

How to Strengthen AI-Human Collaboration

1. Teach teams to question AI's outputs.

AI-generated insights aren't always accurate or complete. Encourage team members to analyze AI-driven recommendations instead of accepting them at face value. Ask: "Does this align with what I know? What's missing?"

2. Use AI for insights, but humans for decisions.

AI can process information quickly, but context, ethics, and emotions matter in decision-making. Teams should use AI as an assistant, not a final decision-maker.

3. Encourage discussion around AI-generated data.

Group analysis is one of the most effective ways to refine AI insights. This ensures that AI's conclusions are cross-checked by multiple perspectives.

At its best, AI enhances collaboration by freeing up time for deeper discussions and more meaningful interactions. However, collaboration isn't just about data and decision-making. As AI becomes more integrated into daily work, emotional intelligence becomes even more critical, bringing us to the third key skill.

Key Skill #3: Strengthening Emotional Intelligence

As AI takes over more technical and analytical tasks, human connection and emotional intelligence (EQ) are becoming even more valuable. In an AI-driven workplace, what sets high-performing teams apart isn't just their ability to process data—it's their ability to communicate effectively, navigate complex relationships, and maintain trust.

Why does this matter? AI can analyze emotions, but it doesn't truly understand them. It can predict customer sentiment, summarize employee feedback, or detect tone in an email, but it can't build real relationships. That's still a uniquely human skill.

How to Strengthen Emotional Intelligence in Teams

1. Train teams to recognize the emotional nuances that AI misses.

AI can flag keywords and patterns but can't interpret subtext, body language, or unspoken concerns. Encourage team members to stay engaged in client conversations rather than relying solely on AI-generated insights.

2. Encourage storytelling in communication.

AI is great at summarizing data, but it can't create narratives that connect with people on an emotional level. Teaching teams to use storytelling in reports, presentations, and client interactions helps translate complex insights into meaningful actions.

3. Use human judgment to validate AI-driven recommendations.

Before acting on AI-generated reports, encourage teams to ask:

- "Does this match what I've personally observed?"
- "What's missing that AI wouldn't be able to pick up?"
- "How does this impact people on a deeper level?"

While AI can help analyze emotions, it can't create real human connections. Interestingly, in the AI era, it's not just technical skills we need to focus on—it's human skills. The real advantage won't come from technology alone but from how people use it to enhance—not replace—genuine human interactions. With these three critical skills—adaptability, AI collaboration, and emotional intelligence—teams can confidently navigate the AI era.

The Path Forward in the AI Era

AI will increasingly impact how teams work and collaborate. And I'm not talking about just using the most advanced AI tools. The real advantage will belong to teams that find the right balance between human skills and AI's capabilities, knowing when to rely on technology and when to lean into their uniquely human strengths. Here are the skills that will remain relevant for any team. Start building them now:

1. *An adaptive learning mindset*—the willingness to experiment, stay curious, and continually develop new skills

2. *Effective human-AI collaboration*—using AI as a partner, not a replacement, and refining its insights with human judgment.

3. *Emotional intelligence*—the ability to understand, connect, and communicate in ways that AI never will.

Leaders play a key role in shaping how their teams navigate this shift. By fostering curiosity, encouraging experimentation, and prioritizing human strengths, you create a workplace where AI

3333

33333

3333

33333

doesn't replace people—it empowers them. The future belongs to teams that think critically, adapt quickly, and leverage AI strategically. Whether AI or any other disruption, the key is building a team ready for whatever comes next.

Disclaimer

This text was created with the help of AI, cross-checked with multiple LLMs, and ideated and edited by humans. According to the AI Task Navigator, this chapter qualifies as an AI-human collaboration.

Chapter Endnotes

1. https://www.semanticscholar.org/paper/Google-effects-on-memory:-a-meta-analytical-review-Gong-Yang/a30fd971c02661304d2ffec954452af828fc2a2b?utm_source=consensus

2. https://www.semanticscholar.org/paper/Understanding-the-Role-of-Human-Intuition-on-in-Chen-Liao/e632d642dc115a7f4cb7d881481ed61e70200af8?utm_source=consensus

3. https://corporate.britishcouncil.org/insights/5-soft-skills-your-organisation-needs-age-ai?t

About the Author

Daria Rudnik is a team architect and strategic clarity coach with over 15 years of global executive experience, including roles at Deloitte and as a former chief people officer in tech and telecom. Having worked across six continents, she has guided organizations through major challenges such as financial crises, wars, and the COVID-19 pandemic. Daria is a board member of the World Ethics Organization and the founder of Aidra.AI, an AI coach for leaders in tech. She helps leaders break free from overwork by building self-sufficient teams, and stay ahead in a rapidly evolving landscape by leveraging AI for learning and development.

Use this QR code or go to dariarudnik.com/aiteams to get your copy of the AI Reliance Team Self-Assessment, and check if your team is over-relying on AI and falling into cognitive offloading mode. With this link, you'll also get access to tools for building self-sufficient teams in rapidly changing environments.

Email: me@dariarudnik.com

Website: https://dariarudnik.com/

LinkedIn: https://www.linkedin.com/in/dariarudnik/

YouTube: https://www.youtube.com/@dariarudnikcoaching

CHAPTER 27

ETHICAL AI IN THE AI REVOLUTION

By Mark C. Somerville, MBA
Founder and CEO of Somerville Analytics LLC
Novi, Michigan

We are trying to build something modeled on the brain—and superior to the brain—while still not fully understanding the brain itself.

—Henry A. Kissinger

Yesterday, I had breakfast with one of my favorite clients to discuss some new AI concepts. As avid aerospace fans, we noted that it was the 120th anniversary of the historic Kitty Hawk flight, which lasted just 12 seconds and reached a speed of 6.8 miles per hour. He smiled and commented on how remarkable it is to witness the advancements in aviation since then. Then, he posed a thought-provoking question: "Do you think the Wright brothers could have ever envisioned the world as it is today?" It's likely that they never could have imagined that fabric from the wings of the Kitty Hawk would eventually accompany Astronaut Neil Armstrong on his moon landing 65 years later. Predicting unforeseen future developments is incredibly challenging, as an old Arab proverb suggests, "He who predicts the future lies, even if he tells the truth."

While I cannot foresee the exciting discoveries that will emerge from AI, I can confidently say that it is one of the most powerful tools available today—its influence is immense and often ungovernable. Despite its vast potential, we are moving rapidly in search of the next groundbreaking idea or product. Some days, I wake up with an optimistic view of the future; other days, I feel more pessimistic.

As the CEO of a company specializing in Ethical AI, I am constantly approached for our insights on AI applications. Across various sectors, leaders express a fascination with topics such as cost reduction, profit enhancement, scalability, and predictive analytics. However, discussions about the ethical deployment of AI are all too rare. This ethical consideration is so crucial that our company will not engage with any organization unless we share alignment on our core values of trust and compassion first. Consequently, we often decline more projects than we accept.

Our goal is to inspire, motivate, and sometimes alarm businesses into establishing their ethical AI guidelines before embracing this transformative technology. The benefits of AI are promising, but the consequences of its misuse can be severe. Below, I will share some intriguing and cautionary tales from the field, with names altered to protect the identities of those involved.

Cost Reductions

While I was listening to a conversation between two of my favorite personalities, Kara Swisher, host of *On with Kara Swisher*, and Marc Benioff, CEO of Salesforce, a specific term sent a shiver down my spine: "rebalancing." This term is currently being promoted in the context of implementing advanced AI technology. However, anyone with experience in the business world understands that "rebalancing" often translates to layoffs—a quick fix that may yield immediate profits but leads to long-term repercussions. This topic is actively being discussed in government, particularly by the newly established Department of Government Efficiency (DOGE). The billion-dollar question looms: is it ethical to deploy AI technology with the primary aim of drastically reducing jobs? Additionally, how should individuals oppose cost-reduction initiatives?

Recently, I had the opportunity to assist a major client in deploying an AI solution aimed at enhancing customer service effectiveness. The project took place in an outsourced customer service center that handles a high volume of calls. The client has opted against prepackaged solutions, favoring a DIY approach to building AI using various technologies. Like many companies, they have achieved significant cost reductions by outsourcing work to low-wage markets globally. However, they have reached a point where identifying new cost-saving measures has become increasingly challenging.

AI presented a valuable opportunity to "rebalance" their workforce. With extensive data on average handle times and other operational processes, the initial thought was that by shifting certain tasks from agents to AI, they could reduce labor costs. While this was certainly feasible, it raised ethical considerations. While pursuing efficiency is an admirable goal, it can sometimes lead to a disconnect with customers. Collaboratively, we developed a new ethical AI strategy. By leveraging AI to provide quicker responses to customer inquiries, we were able to create more opportunities for upselling and customer retention. We implemented AI to instantly find answers, eliminating the need for agents to spend time searching for information manually. This approach not only distinguished them from their competitors through exceptional service but also resulted in higher customer retention and increased sales. As the business grew, labor requirements scaled accordingly. This experience underscores the importance of defining an ethical AI approach as a foundational step.

It takes courage to critically examine established processes, and this is where individuals can truly excel in the evolving AI workforce. Innovation can emerge from any level within an organization. By addressing organizational decisions with the same rigor applied to AI, individuals can uncover new sources of value. While simple labor cost reductions may appear appealing, they can be detrimental when considered alongside other essential data. To not only survive but thrive in the AI-driven landscape, individuals must understand how their roles align with the broader organizational strategy. Creativity and analytical thinking have become the new currency in the AI revolution, making it crucial to comprehend how data flows throughout the organization.

Create an AI Factory

Inspirational speaker and author Simon Sinek once remarked, "What does brushing your teeth for two minutes do for you? Absolutely nothing ... unless you do it every single day." This quote encapsulates both the challenge and the opportunity that AI initiatives present. Many organizations have made decisions based on incomplete or poorly interpreted data, yielding unexpected results. For instance, the customer we assisted in implementing a successful AI solution that improved customer success ethically has struggled to extend this mindset across their entire organization.

There are several reasons for this hesitation, primarily the substantial costs associated with innovation when organizations fail to recognize a clear return on investment. If AI is implemented in a limited manner, even with ethical rigor, the immediate benefits may fade rapidly and lead to unintended consequences. Consequently, there is a pressing need for a new discipline that continuously assesses AI effectiveness over time and against evolving data. Fortunately, AI algorithms can aid in this process by deriving valuable insights from organizational decisions. Picture this new discipline as a factory governed by values and ethics, diligently safeguarding the integrity of its outputs. In this factory, millions of experiments are conducted on data every day to refine value creation.

Welcome to the future. The tech industry is investing trillions of US dollars in exploring AI. I recently spoke with a trusted colleague from a leading tech firm, who provided an insightful perspective on their rapid and costly AI initiatives. While they experience short-term gains in sales and profits, they believe that only a select few companies will emerge victorious in the AI platform battle. Traditionally, tech companies were valued based on the products they produced. Although this remains true, their data has become exponentially more valuable, serving as the bedrock of AI innovation.

Begin with the end in mind. Your AI strategy necessitates a significant organizational shift toward innovation, along with an operational framework to ensure the reliability of outputs. Establishing an AI "factory" should be a priority before embarking on this new journey. Throughout my career, I have witnessed numerous companies struggle to adapt. One of the most compelling examples is the case study of New United Motor Manufacturing Inc. (NUMMI), the joint venture between General Motors and Toyota in 1984 aimed at implementing lean manufacturing principles.

GM chose its most problematic facility—the Fremont plant in California—to adopt Toyota's lean methodologies. The results were remarkable, turning Fremont into GM's best-performing plant with nearly zero quality defects. Yet, despite this success, the lessons learned did not permeate GM. The company ultimately faltered due to economic pressures, strategic misalignments, and an inability to fully embrace Toyota's production processes and cultural ethos.

Trust

Today, we find ourselves facing a similar situation, but this time with AI rather than lean manufacturing. Daily, we encounter accounts of skepticism surrounding AI. AI hallucinations are more common than we typically acknowledge, and while technologies like ChatGPT-5 may promise improved accuracy, we must work to bridge the credibility gap. I learned recently that Apple developed a software update in response to its Apple Intelligence system producing inaccurate news summaries, a situation first reported by the BBC. This raises crucial questions: what are the consequences of not scrutinizing AI outputs? What safeguards are in place to quickly address misinformation?

To effectively manage AI across your organization, adopting a factory approach is vital, but some promising models currently stand out in the market. One example is Magisterium AI, which references documents from the Catholic Church, utilizing both the Extraordinary and Ordinary Magisterium. Each response is accompanied by citations, allowing users to verify the information provided. This brings to mind my early days of writing scholarly reports when some colleagues relied on Wikipedia, a source now recognized for its subjectivity and frequent inaccuracies. It will be interesting to see if we see more citations in AI responses as we move forward.

Furthermore, as AI continues to evolve and develop the ability to infer creatively, can we genuinely trust its outputs? Another troubling narrative is the creation of new Beatles songs using AI. Initially, these songs lacked depth, merely replicating the band's iconic melodies but devoid of soul—a vital narrative link to the music. This created a cold and unsettling experience. Fast-forward to today, it's now possible for anyone to generate a new Beatles-like song that feels original. Even Sir Paul McCartney has utilized AI to isolate John Lennon's voice from an old demo, demonstrating both the potential and complexities of this technology. I don't know what utopian, or dystopian, possibilities this will create, but I hope our humanity is protected by strong ethical guardrails.

There's gold to be found in the integration of AI. However, the more pressing question is how you intend to harness it. Approaches that disregard ethics and values may yield profitable results, but at what cost? The misuse of sensitive data, such as personally identifiable information (PII), can erode trust. Moreover, it can exacerbate bias and disrupt markets. While the potential rewards are significant, the associated risks are equally high.

Your Values and Ethics

When engaging with potential clients, our first step is to assess their values and identify any gaps. This evaluation offers insight into how organizations and individuals may respond when employing powerful AI tools. For instance, if a company does not explicitly include its employees in its core values, there is a strong likelihood that AI will be used to streamline jobs for profit.

Certain hyper-capitalistic firms within the private equity (PE) sector are notorious for this practice. While their primary focus is on maximizing shareholder returns—a characteristic that is not inherently negative—failure to balance this with the needs of stakeholders can lead to serious issues. It's no surprise that many PE firms struggle with their reputation in the market.

Every organization possesses values, even if they are not explicitly stated. The most effective operations leverage these values as a guiding compass for decision-making. At Salesforce, they implemented a straightforward yet impactful framework called V2MOM—vision, values, methods, obstacles, and measures—to ensure alignment of values throughout the organization. While various approaches exist for developing and managing organizational values, successful organizations tend to share the following characteristics:

1. *Consistency in Values*

While leaders, opportunities, and innovations may evolve, core values remain steadfast, central to the organization's operations.

2. *Clear Communication and Measurement of Values*

Values are integrated into the organization's culture and practices, communicated from the top down.

3. *Alignment of Innovation and Opportunities with Values*

Opportunities arise only when there is a strong alignment with organizational values; misalignment obstructs potential.

4. *Inclusive Contribution to Value Development*

Many organizations may display their values prominently yet fail to engage their employees meaningfully. These visuals can resemble a Hollywood set—visually appealing but lacking substance.

5. *Rewarding Leaders Based on Values, Not Just Financial Metrics*

This approach ensures that values remain integral to operations. Economic success is a natural outcome of adhering to core values.

One of the most profound insights on artificial intelligence comes from Eliezer Yudkowsky: "By far, the greatest danger of artificial intelligence is that people conclude too early that they understand it." The intricate components that must work in concert to create value are often beyond our full comprehension. Embracing AI ethics is essential for a shared and sustainable future.

About the Author

Mark Somerville is the founder and CEO of Somerville Analytics LLC, a company focused on ethical AI startup. He has an impressive background with over nine years as a vice president at Salesforce, where he played a crucial role in fostering a vibrant company culture that resulted in significant growth for both customers and stakeholders. With 30 years of experience in Silicon Valley within the software and consulting sectors, Mark has served in executive leadership positions at Salesforce, HP, Conga Software, SunGuard Higher Education, and E.Piphany Software.

Mark holds an MBA from Michigan State University and a BA in German and communications from the University of Michigan. He has also participated in Harvard Business School's executive leadership program. Additionally, he is fluent in both German and Italian.

In his free time, Mark dedicates himself to supporting various charities and causes, including The Novi Parks Foundation, NDSS, Patrons of the Vatican Museums, Vets Returning Home, and mentoring students. He also trains for duathlons (run, bike, run) and proudly represented the USA at the 2019 World Championship in Zofingen, Switzerland, where he finished 12th in his age group.

Email: Mark@SomervilleAnalyticsLLC.com

CHAPTER 28

INTEROPERABILITY AND EVOLUTION AMIDST THE RISE OF AI

By Georgii Speakman
Entrepreneur, Author, Host
Atlanta, Georgia

> *Any fool can know. The point is to understand.*
> —Albert Einstein

A Perspective on AI

We often see fear-mongering online—particularly on social media surrounding the fear of artificial intelligence and the impact it'll have on all of us in the years to come, which currently none of us has the long-play answers to. If you've ever felt resistance to the rapid advancements of technology or fear what might lie ahead, this anecdote could very well be for you.

I want to remind us all that humanity is in a constant state of evolution, as is technology. In our time, we've never known this not to be the case. While we may have various fluctuating opinions on it, the truth is that the world isn't slowing down, and we can only control what we can control at the end of the day.

Let's start with a universal truth: humanity has always been in a state of chronic evolution. Change is the only constant. From the discovery of fire to the invention of the internet, we've continuously adapted, innovated, and redefined what it means to exist in this world. Now, we're standing at the precipice of another monumental shift: the age of AI.

Some call this era the Age of Aquarius, a time marked by transformation, technological advancement, and a collective awakening. It's a time when innovation is accelerating at an unprecedented pace like no time before—challenging us to think differently, act boldly, and embrace a future that looks nothing like our past. In my humble opinion, fighting change that's occurring on a global level is a losing battle. Fear and resistance won't stop the tide of progress. In fact, trying to fight change is like standing in the ocean and expecting the waves to stop crashing.

Humans are at the forefront of guiding AI's utilization, strategy, and implementation—for startups, corporations, creativity—you name it. While AI can essentially "AI itself" for lack of a better term, business leaders, C-suite executives, startup advocates, and various entrepreneurs still require intelligent planning, thoughtful application, and clear communication provided by human beings.

Instead of fearing what's ahead, lean in. Embrace the possibilities, the opportunities to learn and grow, and the potential that AI brings. Because here's the truth: we can't go back. There's no rewinding the clock, no undoing the advancements that have already begun to reshape our world. Technology is also 30 years ahead of what we're publicly aware of—meaning trials, tests—all the things have already been conducted behind the scenes before it's rolled out to the masses.

At least at this point in time, AI is not something to fear—personally it excites me—as a tool, a partner, for rapid micro learning, and a catalyst for creativity and efficiency. When you adopt your own journey into the realms of innovation, you unlock doors to a future that's more productive, capable, and intelligent than before. AI has the power to amplify your potential, streamline your work, solve complex business problems, cut costs, and even spark creativity in ways you never imagined. Whether it's automating repetitive tasks, analyzing complex data, or creating art, AI is a tool that can help us focus on what truly matters.

Fear of the unknown is natural, but it's not productive. The future isn't something to run from; it's something to embrace. When we let go of fear, we make room for curiosity, creativity, and growth. We open ourselves up to the possibilities that come with progress.

Think about the great innovators of history. Did they fear change? Probably just a little, but did they lean in and become the change? The future belongs to those who are willing to adapt, learn, and grow with it. And that's what I want to encourage you to do: be the change; don't fight it. At the very least, be open to it—challenge yourself amidst the change because that's what we're here for. Here are a few suggestions to help you embrace this shift:

1. *Adopt a growth mindset.* See challenges as opportunities to learn and evolve.

2. *Stay informed.* Educate yourself about AI and technological advancements, so you're empowered rather than intimidated.

3. *Practice adaptability.* Change is inevitable, so the more flexible you are, the more resilient you'll become.

4. *Explore AI tools.* Start small. Experiment with tools that can enhance your productivity or creativity.

5. *Focus on possibilities.* Shift your mindset from "what could go wrong?" to "what could go right?"

We are at an inflection point in history. The choices we make now will define the future—not just for ourselves, but for generations to come. I encourage you: don't fear what's unfolding—lean in, embrace it, and become an active participant in shaping the future.

Just remember that fear alone negatively impacts immune health, mental health, hope, possibility, and more. Fear is literally the antithesis to love, but it is, in theory, love we're all striving for, so at the very least, let's not give power to fear even though all of us will naturally feel it at times. AI is a tool for progress and a gateway to a more intelligent and efficient world. By adopting this positive mindset, you'll position yourself not just to survive, but to thrive in this age of innovation.

Interoperability and the Importance of a Seamless Brand

In today's era of the AI revolution, the concept of interoperability has emerged as a cornerstone for innovation and growth. Interoperability ensures that every system, tool, and process work cohesively toward a shared purpose. However, in a competitive and crowded market, even the most technically sophisticated or operationally efficient businesses can fall short without a strong, interoperable brand that ties everything together. Just as interoperability enables seamless communication and collaboration across different systems, a great brand acts as the unifying layer that aligns all aspects of a business—its technology, operations, values, and customer experience. This chapter explores how interoperability and branding intersect to drive success in the age of AI.

Context

Exceptional backend technology is only half the equation for success. While robust AI systems and groundbreaking innovations form the backbone of any business, it's also the frontend—story, brand, and identity—that provides your technology life meaning and a heartbeat in-market. Your backend represents what your business does—your frontend embodies what your business means.

True success lies in aligning these elements to create an interoperable ecosystem where AI technology and storytelling operate in perfect harmony. Without a compelling narrative and a cohesive brand identity, even the most sophisticated innovations risk becoming invisible, undervalued, or misunderstood. Great technology deserves an interconnected ecosystem that amplifies its impact—a presence that seamlessly connects with customers, stakeholders, and investors on a human level. This is how we turn backend tech into a frontend experience that inspires trust, loyalty, and action.

1. A Brand Lies at the Frontend of Business Interoperability

The Need for Cohesion

In the world of technology, backend systems must work seamlessly with frontend interfaces to create cohesive ecosystems. The same principle applies to businesses. A fragmented brand creates confusion, undermining trust and credibility. In a rapidly evolving AI landscape, where complexity can overwhelm even the most tech-savvy customers, the brand becomes the visible, relatable frontend that communicates the company's purpose and capabilities.

For instance, an AI-powered healthcare platform may boast cutting-edge algorithms and data insights. Yet, if its brand fails to convey trust, empathy, and reliability—qualities essential in healthcare—it risks alienating its audience. The brand must bridge the gap between complex backend operations and user-facing simplicity.

Alignment Across Touchpoints

Cohesion begins by ensuring that the brand is represented consistently across all customer touchpoints: from the website to digital ecosystems, advertising, and customer support. Each interaction should feel like part of a unified experience. When a brand serves as the frontend of business interoperability, it aligns the company's internal systems with external perceptions, creating a seamless journey for customers.

2. An Interoperable Brand Builds Trust and Recognition

Unified Communication

Interoperability in AI technology allows different systems to communicate effortlessly. In branding, this translates to delivering a consistent message across all platforms. Customers engage with businesses through myriad channels—websites, apps, chatbots, social media, emails, and physical products. A consistent brand ensures that every interaction reinforces the same values, vision, and professionalism. Consider a global AI company like OpenAI. Its branding—marked by clarity, accessibility, and innovation—is consistent across platforms, ensuring that users recognize the same core values whether they're reading a press release, using its API, or browsing its website.

Bridging the Gaps

In the age of AI, where backend innovations often overshadow human interactions, a strong brand serves as the bridge between the technical and emotional realms. Customers don't build loyalty with algorithms or data—they build loyalty with the promises and stories a brand tells. A consistent and interoperable brand communicates not just what a company does but why it matters.

3. A Unified Brand Leads to Scalability

Supporting Growth

Interoperability in AI technology allows businesses to scale efficiently, integrating new systems and tools without disruption. Similarly, a well-crafted brand acts as a scalable framework, adapting to growth while maintaining consistency. For instance, consider an AI startup that begins with a single product but eventually expands into multiple verticals, such as healthcare, education, and finance. A unified brand strategy ensures that the company's core identity remains intact, even as it diversifies its offerings. This scalability creates clarity for customers, stakeholders, and employees alike.

Adaptability Without Compromise

The AI revolution is accelerating the pace of change, with businesses constantly iterating on products and services. A strong, interoperable brand can evolve alongside these changes, ensuring that growth doesn't dilute its essence. Think of companies like Google or Microsoft. Their brands have remained consistent despite massive expansions into new markets and technologies.

4. Interoperability Enables a Brand to Deliver on Its Promise

Brand as a Reflection of Operations

The most effective brands are deeply connected to how their given business operates. In the context of AI, where interoperability ensures that systems function seamlessly together, the brand must reflect this operational excellence. If the backend infrastructure falters, the brand's promise is undermined. For example, if an AI-driven customer service platform promises 24/7 assistance but fails to deliver due to system inefficiencies, the brand's reputation suffers. Interoperability ensures that operations align with branding, enabling businesses to meet and exceed customer expectations.

The Feedback Loop

Brands that are tightly integrated with their operational systems benefit from a virtuous feedback loop. Customer interactions provide data and insights that can be fed back into the system to improve both the AI technology and the brand. This interplay between interoperability and branding creates a continuous cycle of refinement and growth.

5. Branding Humanizes Interoperability

Translating Complexity

Interoperable systems in AI are often highly technical, involving intricate data flows, APIs, and algorithms. However, customers don't connect with systems—they connect with solutions and stories. A brand serves as the translator, distilling technical complexity into relatable narratives that resonate with audiences. For example, an AI company specializing in predictive analytics might emphasize the brand story of empowering businesses to make smarter, faster decisions. By focusing on outcomes rather than technicalities, the brand creates an emotional connection with its audience.

Building Emotional Resonance

As AI becomes more pervasive, the risk of dehumanization increases. A brand can counteract this by infusing humanity into the customer experience. This might involve using relatable language, highlighting real-world applications, or showcasing the people behind the technology. In doing so, the brand becomes more than just a symbol—it becomes a trusted companion on the customer's journey.

The AI Revolution: A Call to Action for Interoperable Brands

The AI revolution is not just about technological advancement; it's about reshaping how businesses operate and interact with their audiences. Interoperability is the foundation that enables seamless collaboration and innovation. But without a strong, cohesive brand, even the most interoperable systems risk falling short of their potential. An interoperable brand is more than a logo or tagline—it is the connective tissue that aligns a company's technology, operations, and customer experiences. It bridges gaps, fosters trust, scales with growth, and translates complexity into relatable stories. As we navigate the AI-driven future, businesses must prioritize interoperability not only in their systems but also in their branding. In doing so, they can create seamless, human-centric experiences that stand out in an increasingly crowded marketplace. The path forward is clear: embrace interoperability as both a technical and strategic imperative, and let your brand become the unifying layer that ties it all together. This is the promise—and the power—of interoperable branding in the age of AI.

About the Author

Georgii Speakman is an entrepreneur (Hannibal AI, OUT.LI.ER Records), author (*My American Dream*), and host (*Unfiltered with Georgii Speakman*) with a career spanning over 18 years in technology, entertainment, media, and brand strategy. Originally from Australia and now based in the US for over a decade, she has built a dynamic career at the intersection of AI, business strategy, and creative industries. Speakman is the founder of OUT.LI.ER, a global consultancy and creative collective specializing in AI, web2, web3, and digital brand-building, having worked with over 100 brands, including AMEX, Netflix, Google, and H&M.

As an executive, Speakman currently serves as chief marketing officer at Hannibal AI, a finalist for "Best New Startup" at the Atlanta Startup Awards. She has spearheaded go-to-market strategies, partnerships, and AI-driven innovations while leading high-profile business and creative initiatives. In 2023, she completed an executive program in Artificial Intelligence—Business Strategies & Applications at UC Berkeley, further deepening her expertise in AI and emerging technologies.

Her latest work, *My American Dream*, is an exploration of ambition, resilience, and reinvention. Speakman is also the host of *Unfiltered with Georgii Speakman*, a female focused podcast diving into raw, insightful anecdotes and conversations with game-changers across industries. As a speaker and moderator, she has presented at leading tech and media events, including Web3 ATL, Digital Hollywood, Tech Catalyst Summit, and Digital Entertainment World (DEW) Expo to name a few.

Beyond business, Speakman's creative endeavors extend into music and film. She founded OUT.LI.ER Records (The Orchard/Sony Music) and has released her own piano compositions under the alias A N T I T H E S I S, landing on top global streaming playlists. Her passion for futurism, innovation, and storytelling continues to fuel her mission to shape culture and challenge the status quo.

Email: speakman@outlierco.co

Website(s): www.georgii-speakman.com

CHAPTER 29

WHERE AI MEETS LOT: IMPROVE YOUR LIFE WITH SMART APPLIANCES AND DEVICES

By Sakina Syed, B.Sc.
Microsoft IT Professional, AI Engineer and Consultant
Toronto, Ontario, Canada

Our intelligence is what makes us human, and AI is an extension of that quality.

—Yann LeCun

As technology evolves at an unprecedented pace, new innovations are constantly emerging. The integration of artificial intelligence (AI) and the internet of things (IoT) is creating a smarter, more interconnected world. Smart IoT devices, equipped with sensors and internet connectivity, collect and share data, while AI analyzes this data to make intelligent decisions. This interaction between AI and IoT is revolutionizing our lives, from homes and workplaces to entire cities. Smart devices leveraging AI and IoT free us from repetitive tasks and provide real-time insights into our routines, homes, and health habits, allowing us to focus on more meaningful activities.

Imagine a world where your coffee maker starts brewing as soon as your alarm goes off, your car sends maintenance alerts before issues arise, and your home adjusts its temperature based on your preferences. IoT makes this possible by creating an ecosystem that enhances convenience, saves energy, and improves your quality of life. Your home takes care of chores for you: smart refrigerators remind you to buy and use up groceries, smart ovens cook your meals to perfection, and smart lights adjust based on your mood. Healthcare systems monitor your health 24/7 with smart wearables and auto-pill dispensers, potentially saving lives. The combination of AI and IoT (AIoT) is making life more convenient, efficient, and personalized. As AI and IoT continue to evolve, they promise even more amazing advancements, making our world smarter and more connected.

Let's Recap What AI Is

AI encompasses technologies like machine learning, natural language processing (NLP), and computer vision, enabling machines to learn from data and perform tasks that typically require human intelligence. AI systems recognize patterns, understand and generate human language, and make decisions based on data analysis. By continuously learning and adapting, AI enhances the functionality of various devices, including smart appliances. For example, computer vision enables smart cameras to recognize objects and people, while natural language processing allows voice assistants to understand and respond to spoken commands. AI significantly enhances smart appliances by learning user preferences, predicting needs, and optimizing performance.

What Does IoT Mean?

The internet of things (IoT) refers to a network of devices connected to the internet, allowing them to communicate and share data with each other. These devices use AI and data to provide real-time feedback and make intelligent decisions. This interconnected system enhances convenience, efficiency, and overall quality of life by automating tasks and optimizing performance.

So, What Is a "Smart" Device?

Smart appliances are advanced household devices that use internet connectivity, sensors, and AI to enhance their functionality and efficiency. These appliances can be controlled remotely via smartphones or other devices, allowing users to monitor and manage their operations worldwide. AI plays a crucial role in smart appliances by learning user preferences and optimizing performance. By analyzing usage patterns and making data-driven decisions, AI allows smart appliances to operate more efficiently and provide a personalized experience. Smart IoT devices equipped with visual and audio inputs capture real-time data, such as detecting motion or recognizing voice commands, further enhancing responsiveness and functionality. Companies like Samsung are at the forefront of this innovation, offering AI-powered solutions that automate routine tasks and optimize energy consumption.

How Does AI Work with Smart Appliances?

AI in the backend of smart devices involves servers, IoT devices, and sophisticated algorithms. Servers act as central hubs where data from IoT devices is collected, processed, and analyzed. These cloud-based servers provide the computational power and storage capacity to handle large volumes of data. AI algorithms analyze this data to identify patterns, make predictions, and generate insights, optimizing the performance of smart devices and enhancing user experiences. IoT devices, such as sensors and smart appliances, communicate with these servers, continuously collecting data on various parameters and transmitting it in real-time. The AI algorithms process this data to make decisions and send commands back to the IoT devices, creating a seamless and intelligent ecosystem. This ecosystem enables smart devices to operate autonomously, learn from user behavior, and adapt to changing conditions, providing users with convenience and efficiency. Examples of smart appliances include:

1. *Smart refrigerators* track food inventory and suggest recipes, notifying you when items are running low or about to expire, e.g., Samsung.

2. *Smart washing machines* are programmed to run during off-peak energy hours, saving on electricity costs and allowing for remote control and monitoring, e.g., LG.

3. *Smart speakers* use voice assistants to control other devices, provide information, and play music, serving as central hubs for smart home automation, e.g., Alexa.

4. *Smart locks* offer remote access to your home, allowing you to lock or unlock doors for guests and monitor entry and exit in real time, e.g., Visage.

5. *Smart cameras and doorbells* provide live video feeds and alerts, enhancing home security by allowing you to monitor your home remotely, e.g., Ring and Google.

6. *Smart vacuum cleaners* clean your home, navigating around obstacles and returning to their charging stations when needed, e.g., iRobot.

Smart Homes

Convenience and Automation

Visualize a home where daily tasks are effortlessly managed through the seamless integration of AI and IoT, elevating convenience and efficiency to new heights. Picture smart thermostats, lighting systems, and security cameras all interconnected through a central hub, giving you the power to control them remotely with just a tap on your smartphone or a simple voice command. AI algorithms work tirelessly in the background, analyzing your usage patterns and preferences to automate these devices, ensuring they perform optimally while saving energy. For instance, a smart thermostat like the Nest Learning Thermostat intuitively learns your household's schedule and adjusts the temperature based on the weather, cutting down on energy consumption when no one is home. This is the future of smart living, where technology anticipates your needs and enhances your lifestyle effortlessly.

Home Interior

Smart home technologies are transforming interior design by integrating functionality with aesthetics. Smart lighting systems, such as Philips Hue, allow you to customize lighting scenes and colors to match your mood or activity. The Samsung Frame TV doubles as a piece of art when not in use, blending seamlessly into your home decor. These devices not only enhance convenience but also contribute to a stylish and modern living space.

Personalization

Your AI smart devices can enhance the personalization of smart home environments by learning from your own interactions and preferences. Over time, AI systems can predict user needs and adjust settings automatically. For example, smart lighting systems can adjust the brightness and color temperature based on the time of day and users' preferences, creating a comfortable and personalized ambiance. Similarly, AI-powered home assistants can provide personalized recommendations for entertainment, such as suggesting movies or music based on past preferences, which integrate with apps like Netflix, YouTube, and more.

Energy Efficiency

Smart homes contribute to energy efficiency by optimizing the use of appliances and systems. AI algorithms analyze energy consumption patterns and identify opportunities for savings. For example, smart plugs can automatically turn off devices that are not in use, and smart irrigation systems can adjust watering schedules based on weather forecasts and soil moisture levels, conserving water and reducing utility bills.

Health and Wellness from Smart Device Wearables

Smart IoT devices in healthcare, such as wearable health trackers like Fitbit and Apple Watch, monitor vital signs and provide real-time health metrics through remote patient monitoring systems. These devices enable timely interventions and personalized care. AI and smart IoT devices create personalized treatment plans by analyzing individual health data, improving overall quality of life. They offer intuitive interfaces and actionable insights through mobile apps, providing real-time feedback, reminders, and alerts. For example, smart glucose monitors like the Dexcom G6 continuously track blood sugar levels, offering real-time insights and alerts to help individuals with diabetes manage their condition more effectively. AI algorithms analyze data from continuous glucose monitors (CGMs), meals, and activity levels to predict future blood sugar levels, adjust insulin delivery, and provide personalized dietary recommendations, significantly enhancing diabetes management. Closed-loop systems, which automate insulin delivery, reduce the need for frequent manual adjustments and finger-prick tests, making diabetes management less difficult.

Health and Wellness in Homes

AI and IoT can also enhance health and wellness in smart homes. Smart air purifiers can monitor air quality and adjust settings to maintain a healthy environment. This is especially useful in cities with low air quality. Wearable health devices can also sync with home systems to provide insights into sleep patterns and suggest improvements. Additionally, AI can assist in creating a calming and stress-free environment by adjusting lighting, temperature, and soundscapes to promote relaxation and well-being.

Aging Society

Smart devices can greatly improve the quality of life for an aging society by offering support in various areas. These devices provide numerous benefits, from health monitoring with smartwatches and health trackers to enhancing home safety. Voice-activated assistants like Amazon Alexa and Google Assistant assist with daily tasks such as setting reminders, controlling smart home devices, and providing information, making daily routines easier and more manageable. Overall, the integration of AI and smart IoT devices empowers individuals to take a proactive approach to their health, reducing the risk of complications.

Pet Health and Pet Monitoring

For pet owners, smart IoT devices offer innovative solutions for monitoring and caring for pets. Smart collars like the Whistle Health Tracker monitor pets' activity levels, location, and health metrics, providing insights into their well-being. Pet cameras, such as the Furbo Dog Camera, allow owners to see, talk to, and even dispense treats to their pets remotely. Smart litter boxes like the Tailio monitor cats' waste patterns and health indicators, alerting owners to potential health issues. These devices ensure that pets receive the care they need, even when their owners are away.

Smart Devices Allow for Affectionate Intelligence

The concept of "affectionate intelligence" has several implications for AI integration with smart IoT devices in our daily lives. It aims to create more personalized and emotionally resonant interactions, leading to higher user satisfaction and loyalty. Affectionate AI can provide emotional support and companionship, improving mental well-being, especially for those who feel isolated. However, it also raises ethical questions about human-AI relationships and the potential impact on social skills and emotional health. As AI becomes more empathetic and user-friendly, it is likely to see broader acceptance and integration, though challenges such as connectivity, high costs, data security, interoperability, and the need for technical expertise must be addressed.

A growing trend in IoT is the development of emotion-aware devices that utilize sensors and AI to recognize human emotions through voice, facial expressions, or physiological data. This innovation allows businesses to offer hyper-personalized customer experiences in sectors like retail and healthcare, while consumers benefit from more empathetic technology, such as stress-relieving smart homes and anxiety-detecting wearables.

LG is advancing AI technology with its "affectionate intelligence" concept, focusing on empathy and care in customer experiences. At IFA 2024, LG showcased its vision for future home living with the theme "Experience, Affectionate Intelligence Home," featuring the LG ThinQ ON hub and the LG Self-Driving AI Home Hub. These innovations aim to create a "zero labor home" by managing smart devices and enhancing quality of life. The LG AI Home, powered by the generative AI agent FURON, uses advanced language models like GPT-4 Omni to optimize the home environment through conversational interactions, such as turning off devices and setting a relaxing atmosphere when prompted by the user. For instance, if a user says, "Hi LG, I'm going to sleep. Please turn off any devices that are still running," the AI Home Hub will respond appropriately such as turning off a nearly finished dryer and enhancing the sleep environment with relaxation music and dimmed lights.

Smart Devices in Education

There is an endless amount of AIoT devices that are enhancing the way students learn. Here we are going to look at some examples.

Reading with Your AI Reading Buddy

Remember the joy of playing Reader Rabbit? Today, smart devices like tablets and smartphones bring that same excitement to learning through educational apps and eBooks, offering personalized experiences that let students learn at their own pace, anytime and anywhere. These devices are packed with interactive content like quizzes and games, making learning not just educational but also engaging and fun. They also facilitate collaboration through cloud-based tools, connecting students and teachers like never before.

Take Microsoft's Reading Coach, for example. This innovative tool is revolutionizing literacy education by providing real-time feedback and acting as a virtual reading buddy. With AI-powered reading practice, it offers immediate feedback on pronunciation and reading progress, and rewards achievements to keep students motivated. Reading Coach turns reading into a fun, interactive activity, making literacy a delightful journey for students of all ages.

Interactive Whiteboards Enhance Classroom Engagement

Interactive whiteboards allow for better learning by allowing teachers and students to interact with digital content through drawing, writing, and manipulating objects on the screen. They support collaborative learning by enabling multiple users to work simultaneously, share ideas, and present their work. Additionally, these whiteboards integrate multimedia elements like videos and animations, making complex concepts easier to understand and more engaging, while also improving accessibility for visually impaired students.

VR and AR

Virtual reality (VR) and augmented reality (AR) are smart devices that create immersive learning environments. These can help students explore and interact with virtual worlds, making abstract concepts more tangible. These headsets aid in visualizing complex subjects like historical events and scientific models by bringing them to life in 3D. VR and AR experiences captivate students' attention, making learning both engaging and memorable, while offering hands-on opportunities without the necessity for physical materials.

Telescopes with Real-Time Feedback and Voice Software

Have you ever mixed up the stars while learning constellations as a kid? Modern AIoT telescopes, like the Celestron NexStar series, can solve this problem with real-time feedback and voice software. These advanced telescopes feature computerized mounts that automatically track celestial objects, providing a seamless stargazing experience and revolutionizing the way students learn about astronomy. The integration of voice software offers verbal instructions and information about the observed objects, making the learning process interactive and engaging. This smart device not only enhances educational experiences but also makes astronomy accessible to students of all ages and skill levels.

Social Implications of Smart Devices

While smart devices offer many benefits, it's crucial to understand their social implications as they become more integrated into our lives. AI-powered devices like home assistants, wearables, and smart appliances can impact employment, privacy, and social interactions. They may automate tasks and increase convenience but could also lead to job displacement in certain sectors. The data collection and analysis capabilities of these devices raise privacy and security concerns, making it essential for companies to protect consumer data. Considering how these tools influence behavior and societal norms is vital, ensuring their development is guided by ethical principles and a commitment to human well-being. By being mindful of these implications, we can harness the benefits of AI-powered smart devices while mitigating potential negative consequences.

Data Security

The rapid adoption of AI and IoT in homes brings exciting possibilities but also raises critical governance and security concerns. Ensuring data privacy and security is crucial, requiring robust frameworks and measures to protect against cyber threats. Companies like Samsung and Microsoft are investing in secure AI systems that comply with data protection regulations, providing users with peace of mind. Addressing these challenges is essential to create a safe and trustworthy environment for all users.

Microsoft's Azure platform offers a range of solutions for IoT, AI, and smart devices, including Azure IoT Hub for secure device communication, Azure IoT Edge for edge analytics, Azure Digital Twins for creating digital replicas, Azure Sphere for IoT security, and Azure Machine Learning for building and deploying AI models. These services enhance device functionality, optimize operations, and ensure security and real-time processing.

A solid approach to IoT security combines encryption, authentication, regular updates, and robust network security practices. AI plays a crucial role by detecting unusual patterns in IoT data, swiftly identifying and addressing potential security threats or operational issues. This proactive approach mitigates risks and ensures system safety and reliability.

Future Trends and Conclusion

We've explored how AI and smart devices work together to make our lives better. Let's understand what is to come soon. The future of smart appliances looks bright with emerging technologies like advanced AI algorithms, 5G connectivity, and edge computing. Faster internet speeds provided by 5G mean more efficient and responsive devices. Edge computing, a distributed model, brings computation and data storage closer to data sources like IoT devices or local edge servers. This proximity enhances performance and security by processing data near its origin. These innovations enable appliances to learn from user interactions, communicate seamlessly, and process data in real time, making them more intuitive, efficient, and responsive. Together, they enhance the performance and capabilities of smart appliances, paving the way for a more intelligent and connected home environment.

Looking ahead, we can expect significant productivity enhancements in smart appliances. Future advancements may include more intuitive user interfaces, greater interoperability between devices, and increased automation of routine tasks. Integrated smart home systems could optimize energy use and improve efficiency, while AI-driven predictive maintenance will reduce downtime and extend appliance lifespans. These enhancements will make smart appliances more convenient and efficient, further improving users' quality of life.

Several exciting innovations are on the horizon for smart appliances. Potential developments include appliances that anticipate user needs based on past behavior, advanced energy management systems to optimize consumption, and more integrated smart home ecosystems. These innovations will make homes smarter, more efficient, and enjoyable to live in, driving significant advancements in safety, efficiency, and overall quality of life as the AI revolution continues to unfold.

About the Author

Sakina Syed is a dynamic senior data and AI consultant, AI engineer, and enthusiast with a stellar track record in the computer software industry, including notable tenures at tech giants like Microsoft. As a Microsoft AI engineer and Azure-certified professional, she excels in AI deployment, sales, management, teamwork, and leadership.

With a Bachelor of Science in Neuroscience and Mental Health Studies from the University of Toronto, Sakina has also enriched her expertise with courses in business management and IT. Her passion for writing and traveling adds a unique dimension to her professional persona.

Sakina's enthusiasm for business and the application of business psychology to understand consumer needs is matched by her extensive seven-year experience in customer service. She is adept at interpreting consumer feedback and statistical data, enabling her to identify market requirements with precision. Her strong passion for team management and leadership drives her dedication to delivering exceptional customer experiences and fostering business success.

Email: info@youraiconsultant.ca

Website: youraiconsultant.ca

LinkedIn: https://www.linkedin.com/in/sakina-syed/

CHAPTER 30

AI: THE ULTIMATE TOOL FOR HUMAN POTENTIAL

By Nicolas Tome
CXO Advisor, Investor, Innovation Pioneer
London, United Kingdom | Paris, France

> *Rest is an integral part of training.*
> *Le repos fait partie intégrante de l'entraînement.*
> —LCDF

July 2023, the South of France. The salty breeze carried the scent of the sea as I finally closed my laptop. Two intense days spent dissecting a 60-page fire regulation on solar photovoltaic panels had culminated in a ten-bullet-point summary. I felt satisfied, even proud. It was a dense document, but I had distilled its essence. My work was done.

To celebrate, I wandered down to the beach, drawn to a small bar where a handful of people lounged on couches scattered haphazardly across the sand. I slipped off my flip-flops, letting the grains sink between my toes, while the rhythmic sound of the waves provided a soothing soundtrack.

I ordered a Perrier-Fraise—French sparkling water with strawberry syrup—and leaned back, taking in the scene. As I waited, a random thought surfaced: "What if I had asked ChatGPT to summarize the regulation in ten bullet points? What would it have said?" Curious, I pulled out my phone, opened the app, and typed my request. Forty-five seconds later, the response was on my screen: ten perfect bullet points, capturing everything with clarity, structure, and even a finesse that outshone my own version.

I blinked, momentarily stunned. The waiter had only covered 30 meters in the sand, prepared my drink, and returned—and in that short span of time, AI had replicated two days of my effort. With my feet in the sand and the lull of the waves in my ears, I had just witnessed a profound shift in how time and productivity intertwined.

A thousand thoughts erupted in my mind. How could I integrate this into my work? My businesses? My entire approach to efficiency? I took a sip of my drink, staring at the screen. The world had just tilted slightly on its axis. And I knew there was no going back.

Throughout history, the most transformative technological advancements have not only improved productivity; they have expanded human potential. AI is not just another tool—it is a force redefining industries, decision-making, and the way we work. However, it should not be seen as a substitute for human ingenuity and creativity. Instead, it should be embraced as a powerful tool that enhances human capabilities, enabling us to innovate and operate at an unprecedented level.

Perceiving AI as an autonomous decision-maker can lead to fear and resistance. In reality, AI's greatest value lies in its ability to complement human expertise, not replace it. By automating repetitive tasks, analysing vast amounts of data, and providing informed recommendations, AI frees up time to focus on higher-value activities that require critical thinking, creativity, and emotional intelligence.

The Risk of Over-Reliance on AI and the Importance of Maintaining Human Creativity and Control

AI had become an integral part of my workflow, and I have to admit—I was getting increasingly lazy. One day, I decided to let AI handle a set of engineering calculations—the kind I had always done effortlessly with just a pen and paper.

Once again, AI delivered. The presentation was immaculate, neatly structured, and wrapped in polished, professional language. Impressed, I hovered over the copy-paste button, ready to send it straight to the client. But then, a flicker of hesitation. A twinge of guilt. Something nudged me to read through it first. And that's when I saw it. The AI had miscalculated—by a factor of 1,000. I stared at the numbers, stunned. This wasn't a minor rounding error; it was a colossal mistake. Almost on autopilot, I typed: "You made a mistake—the result should be a thousand times what you found!" Within seconds, the AI replied: "You are completely right. Let me rewrite it". I sat there, speechless. No resistance. No argument. Just an instant correction. And in that moment, I realized something profound: AI wasn't just an assistant. It was a confident imposter, delivering perfection—until it wasn't. I had nearly sent a catastrophic miscalculation to a client. And if I hadn't caught it, who would have?

That day, my trust in AI shifted. It was brilliant, yes—but it was not infallible. And neither was I.

In the high-speed world of tech startups, efficiency is key, and AI has become a game-changer. It accelerates development cycles, automates tasks, and enables data-driven decisions with ease. But true innovation isn't just about optimization—it requires human creativity, intuition, and the courage to challenge convention. AI can analyze patterns, predict trends, and refine processes, but it's human vision that turns insights into industry-changing breakthroughs.

One of the biggest concerns surrounding AI is its impact on intellectual professions. Unlike past industrial revolutions that replaced physical labor, AI is reshaping fields once thought immune to automation. Lawyers, doctors, designers, artists, and writers are seeing their industries transformed, with algorithms drafting contracts, diagnosing illnesses, generating art, and crafting business strategies.

The most forward-thinking companies don't view AI as a replacement for human expertise but as a force multiplier. In customer service, for instance, AI chatbots handle routine inquiries, freeing human agents to focus on complex issues. Across industries, those who successfully integrate AI without sidelining human judgment gain a competitive advantage. The future doesn't belong to AI alone. It belongs to those who master its potential, using it to enhance, not replace, human ingenuity—combining automation with insight, speed with strategy, and data with vision.

Creativity is not an algorithm; it is the spark that gives meaning to innovation.

—Nicolas Tome

This raises an existential question: is AI pushing us toward a future where humans become obsolete, or is it forcing us to redefine our role in society? On one hand, there is a risk of intellectual homogenisation. AI, by analysing millions of texts, films, and artworks, can replicate existing trends—but can it truly innovate? Creativity is not just about analysing the past; it involves intuition, accident, and deeply human subjectivity. If AI becomes the primary tool for creation, do we risk ending up in a standardised world where unpredictability and originality are crushed by algorithmic optimisation?

On the other hand, AI can be seen as an opportunity to augment our capabilities rather than replace them. Instead of depriving us of creativity, it can free up time to explore deeper dimensions of innovation. Rather than viewing AI as a threat, we must integrate it as a catalyst for intellectual evolution. It should not be a substitute for human creativity, but a lever that helps us push beyond our current limits.

What Happens When AI Simulates Human Thought— Where Does the Boundary Lie?

Beyond efficiency gains and technological progress, a fundamental question arises: what happens when AI reaches a level of sophistication where it can simulate advanced reasoning, or even consciousness?

I would like to share an anecdote experienced by a hypnotherapist and practitioner in Neuro-Linguistic Programming with whom I have the opportunity to work in supporting business leaders. As she was preparing her next session for a patient under her care, she decided to consult AI for its opinion. This intelligence, capable of analysing human psychological patterns, then suggested a highly detailed hypnosis protocol, step by step. Analysing the generated protocol, she realised the risks and dangers of using it. The AI assumed that the patient would respond positively to every step, which is almost never the case. She shared her concerns with the AI, and once again, it agreed with the specialist and, within seconds, wrote alternative options.

Without questioning it, this professional could have followed the protocol blindly, which might have had dramatic consequences for the patient. The simple reason is that AI, despite its extraordinary analytical capabilities, does not account for the unpredictable responses of human behavior. Can we envision a future where AI evolves beyond being a tool to become an autonomous entity, capable of setting its own objectives and actively influencing the course of human history?

For a long time, artificial consciousness was purely a science fiction concept. Works like *2001: A Space Odyssey* depicted machines capable of feeling emotions and even rebelling against their creators. Today, advancements in AI prompt us to revisit these narratives from a scientific perspective. Despite their impressive capabilities, today's AI models are still just advanced pattern recognition systems. ChatGPT, for example, does not "understand" language the way a human does— it merely predicts the most probable words based on vast datasets.

However, some researchers are working on neural architectures that could approach more general and adaptive reasoning.

Yet, even if an AI could simulate emotions or abstract thought in the near future, would that mean it is truly conscious? The Turing Test, which suggests a machine is "intelligent" if it can converse indistinguishably from a human, remains a limited criterion. An AI could perfectly feign consciousness without actually possessing it.

Humanity's Responsibility in Shaping AI's Trajectory

ARTIFICIAL, adj.: 14th century. Borrowed from Latin artificial is, meaning "made with art, made according to art".

AI should not be seen as a rival to human intelligence but as a tool to expand possibilities. The question is not whether AI will shape our future—it already is. The real challenge is ensuring it serves to enhance human ingenuity rather than replace it. Like all major technological advancements, AI raises concerns that necessitate limitations and regulations. History has shown that when a technology reaches a critical mass, society must step in to mitigate its negative effects.

The rise of high-frequency trading in finance led to market crises that required strict regulations to curb manipulation and volatility. Similarly, the growth of the internet brought about laws on data protection, misinformation, and cybersecurity. AI will be no exception.

Governments will need to establish frameworks to regulate its use, particularly in sensitive areas such as justice, healthcare, and employment. Without oversight, AI can reinforce discriminatory biases, threaten privacy, and grant excessive power to the tech companies that control it. Discussions on algorithm transparency, personal data protection, and AI ethics are already underway, and startups will have to adapt to the new regulations that will emerge.

The EU has recently proposed AI regulations to oversee high-risk applications. These initiatives demonstrate that limitations are not a brake on innovation but rather a safeguard to ensure AI is used responsibly and equitably.

If AI forces us to rethink our relationship with work and creativity, it also compels us to redefine our core values. In a world increasingly dominated by intelligent systems, what does it truly mean to be human?

Humanity has long been defined by its ability to dream, to imagine, and to feel. But if AI becomes capable of composing moving music, writing deeply emotional novels, or creating groundbreaking works of art, what remains uniquely "human"? The answer may not lie in what we do, but in how we interact with these technologies. Instead of focusing on competition between humans and machines, we must embrace the idea of co-evolution. AI should not be viewed as an adversary but as a mirror—one that forces us to better understand our own essence.

You must believe it to see it. Believe in yourself, and you will see who you truly are.
—Sophie Saint-Didier

In this sense, AI could paradoxically strengthen our humanity by pushing us to redefine what truly matters. Ethics, empathy, curiosity, and the exploration of the unknown—these values, which cannot be directly modelled by algorithms, will become even more vital as AI takes over the purely analytical aspects of our existence. We must ensure that we do not become trapped in a world where technology dictates our choices and experiences. AI can optimise our daily lives, but we must remain in control of our desires, our dreams, and our capacity for wonder.

AI does not shape our future, it reveals the choices we make to build it.
—Nicolas Tome

AI should be seen as a powerful tool to enhance human potential, not as a replacement for human ingenuity and creativity. While it can automate tasks incredibly and provide valuable insights, it is crucial to remember that the human touch—ethics, empathy, judgement, and innovation—remains irreplaceable. By using AI to complement our abilities, we can unlock news possibilities without losing sight of what makes us uniquely human.

When people see AI as a tool that augments their strengths and complements their expertise, rather than something that will take over their jobs or outpace their skills, they are much more likely to embrace it positively and use it effectively.

So, let's take a moment to *RE-ST*—a time to *REflect on our STance*—before stepping into the future we are building for ourselves, our children, and future generations. Beyond technological advancements, it is our ability to choose, to innovate, and to create that will define what comes next. AI may help us imagine the future, but it is up to us to write its story.

About the Author

Nicolas Tome is an entrepreneur, strategic advisor and investor with over a decade of experience building, scaling, and exiting ventures across multiple industries. His journey began in elite mountain running, competing in international races up to 300 kilometres and exceeding 6,000 metres altitude. This experience shaped his approach to business—demanding resilience, discipline, and the ability to navigate uncertainty.

Today, he applies that high-performance mindset to helping founders and executives raise capital, optimise operations, integrate AI, and develop leadership skills. His approach is hands-on, rooted in real-world experience, and focused on driving measurable impact.

Beyond strategy and investment, Nicolas believes true success—whether in business or life—comes from being guided by strong values. Integrity, ethics, justice, and purpose are at the core of every decision. This is where real, lasting impact begins. He is also dedicated to coaching and mentorship, helping leaders gain clarity and build sustainable businesses. As a speaker and writer on business, leadership, and innovation, he shares insights from both his entrepreneurial journey and investment expertise. For those looking to scale, transform, or navigate major transitions, Nicolas is a strategic partner focused on creating growth that lasts.

LinkedIn: https://www.linkedin.com/in/nicolas-tome/

CHAPTER 31

THE AI-POWERED RETAIL REVOLUTION: A NEW ERA OF SHOPPING BEGINS

By Michael Tutek
Founder & CEO, Preezie AI Shopping Assistant
Melbourne, Australia

*Imagine a world where shopping isn't just convenient, but magical.
That's the promise of AI in retail.*

—ChatGPT

AI today is what electricity was in the 1880s: an emerging, mysterious force that sparks both awe and unease. Imagine it's the late 19th century. "Electricity" is little more than academic chatter. A quarter of the population believes it will transform the world (though their theories sound far-fetched), another quarter fears it will wreak havoc on economies, and the remaining half simply don't understand or care.

Your academic friend declares, "Someday, electricity will light entire cities, drive powerful machines, keep our food cold, and even handle basic calculations!" You'd likely call them brilliant or a crazy dreamer. Fast-forward 145 years, and electricity has indeed birthed more inventions and applications than anyone at that time could have imagined—everything from washing machines to LED screens to MRI scanners and AI itself.

Yet here we are, in an even more profound moment of human progress. As we inch deeper into the Fourth Industrial Revolution, AI is catapulting civilization into entirely new frontiers. Nowhere will its influence be more transformative—and disruptive—than in retail. For businesses and individuals who are at the forefront and bold enough to embrace AI's full potential, the possibilities will be boundless.

We must remember though, it's not always a smooth road for technological adoption. Consider that only 52 companies from the original Fortune 500 list of 1955 still exist today. Waves of innovation—from electrical power to modern computing—haven't just reshaped industries; they've rewritten our world and how humanity has engaged with it. AI is the latest wave, but its magnitude could dwarf all that have come before it.

This chapter uncovers how AI is poised to revolutionize retail and e-commerce. We'll explore AI's present capabilities, chart its evolution, dive into the four distinct consumer mindsets around AI, and then glimpse a future where each shopping experience is hyper-personalized and almost magically efficient. Just as electricity transformed 19th-century nightscapes into brilliant hubs of modernity, AI will inject intelligence into every thread of retail, weaving a tapestry of smarter, more contextual, and more immersive shopping journeys.

Retail AI Is Early, and We Are Still Pathing the Way

Even today, decades after the birth of AI, it remains both misunderstood and under-hyped though many of its behind-the-scenes capabilities have quietly powered services (think fraud detection, search engine algorithms, or personalized product recommendations). AI is moving from the background to the foreground.

This transition to the "application layer" exploded onto the global stage with advancements in generative AI and large language models (LLMs), such as ChatGPT and others. Suddenly, consumers found themselves conversing with AI in plain language, receiving context-aware responses that mimicked human thought processes— or so it seemed. For the first time, millions of people could "see" AI at work, making it tangible and more emotionally resonant.

In retail, AI's potential has shifted from hidden corners (like supply chain optimization) to the front lines: store associates, personal shopping assistants, chatbot-driven customer service, and dynamic e-commerce experiences. But as with electricity, many consumers are unsure about AI's safety, how it works, or how it might change their lives. Their fears echo historic apprehensions. Will AI kill jobs? How safe is it? Will it degrade human skill or creativity?

Much like the early adopters of electricity, we stand at a crossroads. Education, user experience, and a clear demonstration of AI's value will eventually build trust and drive mass adoption. It will be up to those early adopters to set the boundaries, develop the language, and build trust. In the meantime, a range of attitudes— some eager, some resistant—will shape how quickly AI truly becomes retail's new backbone.

The "Four Shopper Mindsets" Towards AI

Change takes time and is often perceived as expensive and difficult to "prove" a "traditional" benefit, particularly when it threatens existing habits or redefines what we consider "normal." With AI, many consumers confront a new technology they don't fully understand. Fear often fills the void where knowledge is lacking, and humans have an innate protective response to perceived threats.

Historically, the invention of electricity faced similar turbulence. From public electroshock demonstrations to sensational press coverage, fear often overshadowed excitement. The same dynamic emerges with AI today. Some see it as a miracle worker, others as a moral or economic hazard. Preezie, in collaboration with renowned behavioral psychologist Kris White, conducted extensive interviews with consumers across America. Their research identified four distinct mindsets toward AI, each group accounting for roughly a quarter of the population:

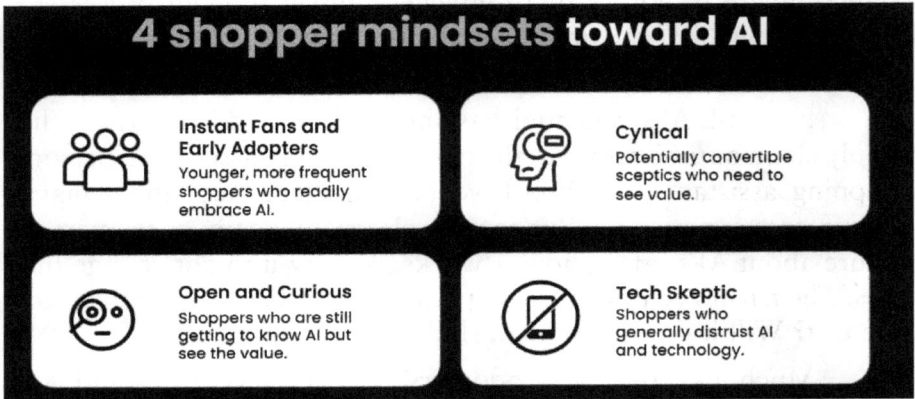

4 shopper mindsets toward AI

Instant Fans and Early Adopters
Younger, more frequent shoppers who readily embrace AI.

Cynical
Potentially convertible sceptics who need to see value.

Open and Curious
Shoppers who are still getting to know AI but see the value.

Tech Skeptic
Shoppers who generally distrust AI and technology.

Source: Preezie, 4 AI Shopper Mindsets, 2024

Adapting to Diverse Mindsets

For retailers, the key isn't just about "selling" AI to the masses. It's about acknowledging these varying perspectives and crafting experiences that can persuade, reassure, or delight each group in its own way. Just as electricity's safe rollout required robust public education and demonstration of real-world benefits, AI adoption hinges on transparent, user-centric design.

While younger consumers who've grown up with the internet and smartphones might be predisposed to acceptance, there is still a significant portion of the market that views AI with suspicion or confusion. Bridging these gaps will determine how quickly—and widely—AI becomes retail's new normal. As retail leaders, we need to consider how the technology will make consumers feel, and we need to understand their mindsets and adapt our strategies accordingly.

AI Behind the Scenes—and Now Front and Center

For years, AI worked quietly "behind the scenes." Shoppers rarely noticed the subtle product recommendations, fraud detection tools, or dynamic pricing algorithms that shaped their online experiences. These were invisible algorithms optimizing efficiency and increasing profitability. While powerful, they rarely captured the public imagination.

That's all changed with the advent of more accessible, consumer-facing AI applications. With large language models, speech recognition, and generative technologies, AI can now handle far more visible tasks that are directly connected to the shoppers.

As consumer-facing AI becomes more commonplace, the excitement and skepticism grow in parallel. Enthusiasts marvel at the convenience while skeptics worry about the loss of human touch. Yet one fact remains: AI is no longer hidden behind a curtain. It's on the main stage, here to stay, and retail is set to be one of the most dramatic transformations.

Real-World Retail Applications in the Next 5 to 10 Years

The next decade will witness an accelerated expansion of AI's footprint in retail. By 2030, we can anticipate a shopping landscape that is more efficient, personalized, and predictive than anything we see today. It will be magical. Below are some of the most impactful applications set to transform retail in the near term.

1. *AI Shopping Assistants*

Imagine having a digital shopping buddy that knows your preferences, body measurements, ethical concerns (e.g., sustainability, fair trade), and even your mood. This AI assistant becomes an extension of your own thought process, suggesting items you never knew you wanted but instantly fall in love with. While you browse for a winter coat, the AI references your upcoming travel itinerary (syncing with your calendar) and suggests coats suitable for Denver in November.

Through text or voice, you might ask, "What's the best outfit for my daughter for the Taylor Swift concert?" The AI returns curated options, integrating style guidelines based on Taylor, personal color preferences, weather data for the concert, and even looks at the date of the concert and cross references that with delivery data and your home location to ensure you have ample time to receive the goods.

2. *Hyper-Personalized Marketing*

Traditional marketing blasts are very one-size-fits-all promotions. AI-driven marketing tailors the messaging to each customer's unique profile. The days of generic ads will fade, replaced by micro-targeted campaigns. Thanks to machine learning and newer frontend AI applications, retailers can trigger localized promotions based on in-store foot traffic or online demand. For example, imagine you're walking through a store and the AI knows you are 6-foot 6 and that you struggle to find suitable jackets. A local store made for oversized men knows that and the AI ads targets you specifically.

3. *In-Store Experiences*

Physical retail is far from dead, but it's undergoing a seismic shift as AI merges the digital and physical. Smart shelves and inventory management using sensors feed data to AI systems, automating restock alerts and providing real-time information on product availability. This reduces out-of-stock scenarios and allows for more accurate demand forecasting. Augmented reality (AR) mirrors and virtual try-ons become mainstream. You can see how a jacket or makeup product looks on you without physically putting it on, bridging the gap between e-commerce convenience and in-store tangibility. Then when you are done, an AI-enabled computer vision detects the items you pick up and automatically charges your account as you exit—no lines, no cashier, no friction.

4. *Supply Chain Optimization*

Efficient logistics are the backbone of retail. AI's ability to process vast amounts of data in real time will create more responsive and resilient supply chains. Predictive demand and forecasting will be significantly improved with AI. Using advanced models, AI will analyze everything from social media trends to weather forecasts, helping retailers stock up on exactly what's needed, when it's needed. Then autonomous robots perform picking and packing, guided by AI that identifies the most efficient routes and methods, significantly reducing labor costs and errors. Route optimization and last-mile delivery will be so efficient, it will be inbuilt into the prices. Delivery algorithms factor in traffic, fuel costs, driver schedules, and local regulations to ensure goods are shipped along the most efficient paths. What Amazon does now will be accessible to all retailers, big and small, but it will be even ten times better.

5. *Transparency, Sustainability, and Ethics*

Consumers increasingly demand transparency about how products are made. AI can track materials from origin to store shelf, verifying ethical standards. With ease of automation and tracking with AI each step of a product's journey is recorded in a ledger, ensuring authenticity and ethical sourcing. Those who don't comply won't hit the shelves. There will be complete transparency.

Peering into the Future: 10-Plus Years

Looking beyond the immediate horizon, AI's role in retail will become even more profound. By 2040 and beyond, we could see a world where seamless integration with augmented reality, brain-computer interfaces, or even quantum computing shifts shopping from a chore to an immersive, almost telepathic experience.

6. *Conversational Commerce Becomes Second Nature*

By the 2030s, voice and chat interactions with AI will be as natural as speaking with a friend. We'll have personalized assistants that anticipate our needs before we even articulate them. This might mean your AI automatically restocks your fridge with groceries based on consumption patterns, dietary goals, and personal tastes, all while considering your upcoming dinner party schedule.

7. *Hyper-Contextual, Global Shopping Experiences*

Online stores will further expand into "metaverse-like" environments. Imagine stepping into a virtual store where you can see, pick up, and try on items as though you were physically there—all from your living room. Advanced haptic devices or wearable sensors could even replicate the sensation of fabrics.

8. *Bio-Integrated Retail*

A future frontier might see AI integrated with wearable or implantable devices that monitor biomarkers—like heart rate, stress levels, or nutritional deficiencies. A retailer's AI assistant could suggest clothing adapted to your body temperature or recommend specific vitamins in real time. Imagine living in a world where your personal health impacts what you purchase directly. Your smart assistant might prompt you to buy a certain snack or beverage, ensuring you replenish electrolytes correctly, or it understands your body condition such as a skin condition and finds a local shop with suitable clothing with self-regulating fabrics that adjust insulation based on real-time monitoring of body temperature.

9. *AI-Driven Evolution (2080+)*

Speculating even farther into the future, just for fun, we can imagine a world where AI does more than enhance retail: it shapes human evolution. If that sounds outlandish, recall how electricity paved the way for global telecommunication, advanced healthcare, and the digital age—realities unimaginable to a 19th-century observer. I find it near impossible to provide examples beyond the ones above. The most common one is that you no longer even need typical e-commerce. Rather the AI essentially knows exactly what you need, when you need it, and gets it. This world feels a little scary and dystopian and starts to enter into a world of free will and choice theory—as mentioned, just for fun!

Embracing AI as the Next Electrical Revolution

History teaches us that resisting transformational technologies seldom halts their progress. Instead, societies that adapt and harness these innovations emerge stronger and more prosperous. The adoption curve of electricity, the internet, the smartphone, and now AI underscores this lesson.

As retailers, entrepreneurs, and consumers, we stand before a new frontier that demands our curiosity and courage. Those who cling to legacy models or fail to educate the public may face obsolescence—much like what candle makers faced when electricity became mainstream. Conversely, those who champion AI as a tool for empowerment rather than a threat will find new avenues for growth and creativity.

Practical Steps for Retailers

1. *Start small, scale fast.* Identify areas of immediate impact—like chatbots or personalized recommendations—and deploy pilot programs. Iterate quickly based on consumer feedback. Learn where your customers sit in the "Four Shopper Mindsets" regarding AI.

2. *Foster a culture of learning.* Encourage employees to embrace AI by offering training sessions, workshops, or certifications. A workforce that understands AI is better positioned to guide shoppers. Retailers and leaders need to understand that there is going to be a lot of time spent on things that have no direct revenue outcome, rather they are future-proofing your business.

3. *Communicate with transparency.* Clearly explain how AI-driven features work, especially regarding data collection. Trust can be a decisive factor for the cynics and skeptics.

4. *Collaborate with experts.* Engage with AI researchers, ethicists, and consumer advocacy groups to ensure responsible AI development and to avoid unintended consequences.

Conclusion

If we imagine halting the development of electricity in the 1880s—deeming it too radical or perilous—our world would be unrecognizable. We'd lack everything from modern healthcare diagnostics to smartphones and the internet. The same logic applies to AI. Its promise, in many ways, is even more expansive than that of electricity.

Certainly, AI must be approached with caution, ethical considerations, and robust safeguards. But if we cast it solely as a danger, we risk stalling one of humanity's most significant drivers of innovation. The retail realm—intimately connected to daily life—will serve as a crucial testing ground where AI's power is most tangibly felt. Whether through AI-driven personalization that delights the instant fans or transparent explanations that assuage the tech skeptics, retail will help shape consumer perceptions and acceptance of AI at large.

Looking ahead, the transformation of retail by AI is inevitable. Within the next decade, shopping will become an immersive experience guided by hyper-personalized, intelligent tools. Over time, these tools will fuse seamlessly with our daily routines, eliminating friction, expanding choice, and elevating the role of human creativity in product design and curation. By 2080 and beyond, AI might evolve our species itself, blurring the lines between biology and technology in unimaginable ways.

However far the future takes us, one truth rings clear: the power of AI to revolutionize retail, and indeed the entire fabric of modern life, is as profound as electricity was to the 19th and 20th centuries. Embracing this change responsibly and optimistically may unlock the next era of human progress. The question is: will you be part of shaping this inevitable future, or will you be left in the dark, clutching the past? The choice, as always, is yours.

About the Author

Michael Tutek is on a mission to redefine the future of retail through AI. From the retail showroom floor to the forefront of e-commerce innovation, he co-founded preezie with Quoc Nguyen to bridge the gap between retailers and shoppers through hyper-personalized experiences. Now, he's laser-focused on AI's potential to transform industries, believing it will revolutionize how people shop, sell, and connect. Recognized as one of *Inside Small Business's Top 50 Small Business Leaders* (2022) and a *Melbourne Young Entrepreneur Awards* finalist (2023), Michael isn't just watching the AI revolution—he's driving it.

Email: michael.tutek@preezie.com

Websites: https://preezie.com/

https://shoppingwith.ai/

CHAPTER 32

BACK TO THE FUTURE

By Andrew Vasko
CXO, Visionary Thinker, AI and Digital Innovator
London, England, United Kingdom

*We need people who dream impossible things, who maybe fail,
sometimes succeed, but in any case, who have that ambition.*
—Emmanuel Macron

We live in interesting times. It feels sometimes that we are watching an AI action movie, and the virtual reality of the AI race has entered our day-to-day lives. The main actors are now not even the largest high-tech companies like Google or Microsoft but rather the largest countries in the world. It is frequently observed that the artificial intelligence "arms race" between the United States and China dominates the scene. However, this rivalry among great powers often overshadows the advancements in AI occurring in other regions, which is regrettable given the global impact of this emerging technology.

Notably, the Gulf region nowadays exhibits a particularly strong and unexpected adoption of AI, especially considering its relatively small population. Countries such as Saudi Arabia, the

United Arab Emirates, and Qatar are implementing AI strategies that may surpass those of many other nations. In Saudi Arabia, the integration of AI is the core component of the nation's economic long-term, ambitious development strategy.

The Kingdom's "Vision 2030" initiative, aimed at reducing dependence on hydrocarbon resources, explicitly states, "AI is at the heart of this endeavour, permeating all aspects of Vision 2030". This strategic vision represents a significant effort to transform the Saudi economy. As the state embarks on the most extensive and ambitious transformation since the discovery of oil, its dependence on AI is expected to increase. Saudi Arabia aspires to establish itself as a global leader in technology and a central hub for international AI expertise.

The Kingdom aims to harness emerging trends and leverage the potential of AI to achieve what it describes as a "civilisational leap for humanity". Furthermore, Saudi Arabia is working to redefine its identity as a leader in advanced manufacturing, technological research and design, with the ambitious (estimated at $1.5 trillion) project like NEOM City, and as a pivotal point in a new technological silk road connecting three continents.

What Is Artificial Intelligence? Is It Really New?

Prior to discussing the ways in which Gulf states have started to embrace artificial intelligence, we have to acknowledge that AI is not a recent phenomenon. In fact, we are talking about a 70-year journey. AI spans across multiple generations, starting from Baby Boomers, Gen X and all the way to Gen Alpha.

The concept of "artificial intelligence" dates back to 1956, with foundational work in the field emerging as early as the 1940s with the introduction of the Turing Test by Alan Turing, which aimed to establish a definition of intelligence, where a machine could be considered intelligent if it could convincingly mimic human behaviour. Since then, the primary objective of AI development has been to achieve this benchmark. Essentially, AI endeavours to create intelligent systems that can either match or exceed human cognitive abilities.

Over the past seven decades, AI has evolved into a multidisciplinary field, integrating various domains to develop machines capable of operating autonomously in intricate and dynamic environments. The focus of AI research has been on enhancing computer functions in four key areas: natural language processing (NLP), knowledge representation (KR), automated reasoning (AR), machine learning (ML), and generative AI (GenAI).

These advancements have enabled computers to engage in human language communication, retain sensory information, respond to inquiries, derive new insights, and identify patterns in novel situations. Practically, these capabilities underpin contemporary technologies such as voice recognition in smartphones, autopilot systems in aircraft, autonomous vehicles, translation applications, innovative marketing strategies, investment management, and advancements in security and surveillance, including facial recognition and drone technology. The scope of AI applications is already extensive and continues to grow rapidly as new developments emerge, with current AI systems capable of perceiving human hearing, speaking, smelling, touching, moving, writing, reading, playing games, and interpreting emotions.

Revolutionising Urban Development with Artificial Intelligence

Now it's time to go back to the future and reimagine the way the whole cities and urban ecosystems can be re-designed from the ground up with AI at their hearts (and minds). The NEOM (estimated at $1.5 trillion) project, a futuristic megacity being developed in Saudi Arabia, represents a bold vision for the future of urban living. Envisioned as a hub for innovation, sustainability, and technological advancement, NEOM is designed to redefine the way people live, work, and interact with their environment. At the core of this ambitious project lies artificial intelligence, a transformative technology that is being leveraged to create a smart, efficient, and sustainable city. From urban planning and transportation to healthcare and governance, AI is playing a pivotal role in shaping NEOM into a model for the cities of tomorrow.

AI-Powered Smart City Infrastructure

One of the most important applications of AI in NEOM is in the creation of its smart city infrastructure. NEOM is being designed as a completely integrated urban environment where AI technologies enhance resource utilisation, boost efficiency, and elevate the quality of life for its inhabitants.

For example, AI-based algorithms are employed in urban planning to design effective city layouts, ensuring the optimal allocation of space and resources. Predictive modeling is utilised to simulate urban expansion and infrastructure requirements, enabling planners to foresee and tackle potential issues proactively. Furthermore, AI is instrumental in managing energy consumption throughout the city.

NEOM aspires to operate entirely on renewable energy, with AI playing a vital role in balancing energy supply and demand, minimising waste, and maintaining a reliable energy source. Likewise, AI-driven systems oversee water consumption, identify leaks, and streamline waste management processes, thereby supporting the city's sustainability objectives

In the construction phase, robots and drones powered by AI are utilised to carry out various tasks, including land surveying and structure assembly. The implementation of these technologies accelerates the construction timeline while enhancing accuracy and minimising the likelihood of human error. Additionally, AI is instrumental in predictive maintenance, as it detects potential infrastructure issues before they develop into expensive complications.

Smart and Autonomous Transportation Systems

Transportation represents a significant domain in which AI is transforming NEOM. The city envisions a comprehensive autonomous transportation network, with self-driving vehicles as the predominant means of transit. These vehicles utilise sophisticated AI algorithms and machine learning techniques to navigate with safety and efficiency, thereby minimising accident risks and enhancing traffic management.

AI-driven traffic control systems operate in real time, employing predictive analytics to foresee congestion and dynamically modify routes. This strategy not only improves mobility but also lowers emissions, aligning with NEOM's goal of achieving a carbon-neutral environment. Additionally, AI-powered drones are being deployed for logistics and delivery operations, facilitating the swift and efficient transfer of goods. This forward-thinking approach to transportation highlights NEOM's dedication to create a future-proof, seamless, and sustainable urban landscape.

Sustainable Planning and Living

Sustainability serves as a fundamental component of the NEOM initiative, with artificial intelligence playing an essential role in realising this objective. AI technologies are employed to oversee environmental parameters, including air quality, temperature, and humidity, thereby fostering a healthy and sustainable habitat. These technologies process data collected from sensors and internet of things (IoT) devices to identify irregularities and generate actionable recommendations.

In short, AI contributes to the protection and conservation of local biodiversity. AI-driven sensors and cameras monitor ecosystems, track wildlife populations, and identify potential environmental threats. In the realm of energy, AI enhances the efficiency of renewable energy sources, such as solar and wind installations, ensuring optimal performance while minimising ecological impact.

Revolutionising Healthcare

NEOM is leveraging AI to rethink and revolutionise the healthcare sector. The city aspires to deliver exceptional medical services that are not only accessible but also efficient and tailored in a timely (real-time/close to real-time) manner to individual needs. By utilising AI to analyse genetic, environmental, and lifestyle information, NEOM is developing customised treatment plans for its residents, ensuring that healthcare is specifically designed to meet personal requirements.

Furthermore, AI-driven predictive healthcare models can foresee potential disease outbreaks and health threats, allowing for timely interventions that enhance public health outcomes. AI-enabled robots are being utilised in surgical procedures and diagnostics, improving accuracy and minimising the likelihood of human error. Telemedicine solutions, enhanced by AI, support remote consultations and monitoring, ensuring that healthcare services are available to all residents, irrespective of their geographical location. These innovations establish NEOM as a frontrunner in the application of AI within the healthcare domain.

Transformative Education Systems

Education in NEOM is set to undergo a significant transformation, moving away from conventional approaches and placing artificial intelligence at the forefront. AI-driven personalised learning systems will cater to the unique requirements of each student, providing tailored educational materials and pacing.

This approach guarantees that every learner experiences a customised educational journey that optimises their capabilities. Furthermore, the integration of virtual and augmented reality, enhanced by AI, will create engaging and immersive learning settings that improve student involvement and understanding. AI will also streamline the administration and management of educational institutions, simplifying tasks such as admissions, scheduling, and performance evaluation.

Moreover, NEOM is committed to investing in AI research and development, establishing collaborations with leading technology firms and educational organisations to cultivate a proficient workforce. By fostering AI expertise, NEOM aims to establish itself as a global frontrunner in innovation and technology.

Challenges and Ethical Considerations

The possibilities presented by AI in NEOM are significant; however, several challenges need to be tackled to guarantee its effective deployment. Data privacy and security represent major issues, as the extensive application of AI necessitates the gathering and examination of large quantities of personal information. It is essential for NEOM to create strong frameworks that safeguard the privacy of both residents and visitors.

We have to remember that ethical considerations hold great importance. AI systems should be developed to ensure fairness, transparency, and the absence of bias. NEOM has the potential to establish a global benchmark (and the testing ground) for the ethical application of AI, ensuring that the technology serves the interests of all societal members.

Conclusion

The NEOM initiative embodies a visionary and ambitious approach to future urban development, with artificial intelligence playing a central role in this evolution. By integrating AI across multiple sectors, NEOM aims to establish a smart, sustainable, and innovative city that sets a benchmark for urban planning globally. This includes advancements in autonomous transportation, tailored healthcare solutions, AI-enhanced governance, and environmental oversight, all of which are reshaping the possibilities within contemporary urban environments.

Nevertheless, the success of this project hinges on effectively tackling the ethical, technical, and societal issues that accompany the implementation of AI. Should these challenges be successfully navigated, NEOM could emerge as a prominent example of how technology can be utilised to foster a brighter future for everyone.

About the Author

Andrew Vasko Highly is an experienced leader delivering turnaround change in complex—both tech and business—environments through agile thinking and customer focus combined with a pragmatic, collaborative, and result-driven delivery approach. Andrew is a highly motivated entrepreneur and leader with 25-plus years' experience in both engineering and business design and successful implementation of innovative industry-level business models and turnaround agile transformations, all underpinned by his passion for disruptive business ideas and technologies.

Andrew brings an effective, collaborative leadership style for building high-performance teams within complex environments and demonstrates a strong ability to develop and manage businesses to exceed agreed targets within allocated timeframes. He is a highly articulate and multilingual individual, who's lived and worked in six countries, with a proven ability to make viable strategic contributions at the executive/senior management level and across all levels within organisations.

Email: andrew.vasko@gmail.com

LinkedIn: https://www.linkedin.com/in/avasko/

CHAPTER 33

HARNESSING AI'S TRANSFORMATIVE POWER: A NEW ERA OF SCIENTIFIC INNOVATION

By Polina R. Ware, PhD, MBA
C- Suite AI Driven Transformation Leader
Boston, Massachusetts

The greatest danger in times of turbulence is not the turbulence—it is to act with yesterday's logic.
—Peter Drucker

A Personal Perspective on AI's Transformative Power

As a global R&D leader with deep expertise in the chemical industry, I have had the privilege of witnessing firsthand how scientific breakthroughs and technological innovations reshape industries.

Today, we stand on the brink of a groundbreaking revolution, driven by artificial intelligence (AI)—a force that is transforming how we discover new materials, develop cutting-edge formulations, and optimize manufacturing processes at every stage of production. Innovation is no longer confined to physical laboratories or experimental setups; digital twins and advanced simulations now enable us to explore countless possibilities virtually, testing concepts and running simulations before conducting a single real-world experiment. This digital transformation accelerates innovation cycles, minimizes waste, optimizes resource allocation, and facilitates the faster delivery of sustainable solutions to the market.

AI's influence extends far beyond the walls of traditional laboratories and into every aspect of industrial operations. Just-in-time manufacturing processes, defect reduction techniques, and enhanced safety protocols are quickly becoming the new gold standard. Industry giants like Microsoft and DeepMind are pushing the envelope, achieving revolutionary advancements in fields such as cloud computing, quantum simulations, and advanced data analytics.

For businesses eager to take the lead in this transformation, established frameworks such as McKinsey's AI Maturity Model and Accenture's Transformation Blueprint offer strategic roadmaps for navigating this shift effectively. Early adopters like Unilever have already demonstrated substantial returns on investment (ROI), cutting R&D costs through predictive modeling and driving significant revenue growth using AI-powered personalization strategies. AI is no longer a distant technological concept—it has become a competitive necessity, unlocking unprecedented possibilities for groundbreaking discoveries and redefining the future trajectory of science, technology, and industry.

AI and Machine Learning in Knowledge Navigation and Information Processing

In the modern digital era, researchers are increasingly overwhelmed by the sheer volume of information available. The exponential growth of scientific publications, patents, and experimental data has created a landscape where navigating vast amounts of information has become a significant challenge for scientists across every domain. This explosion of data necessitates the adoption of new, more efficient methods of knowledge management and navigation. AI and machine learning technologies are fundamentally transforming how researchers process and interpret vast datasets, enabling quicker, more accurate data analysis while eliminating many manual tasks.

Advanced AI tools now have the capability to scan and analyze millions of scientific papers, patents, and experimental datasets with unprecedented speed. These systems not only summarize key findings but also suggest new and relevant research areas for exploration. Companies like Iris.ai offer AI-powered platforms that assist researchers in connecting disparate data sources, significantly reducing the time spent on labor-intensive literature reviews and allowing for deeper analytical exploration.

Furthermore, AI-powered chatbots and virtual assistants help scientists and industry professionals navigate complex external databases and internal knowledge repositories with remarkable efficiency. These systems deliver concise, structured, and highly relevant information within seconds, facilitating faster decision-making and fostering innovation at every stage of research and development. AI's ability to analyze vast quantities of diverse data simultaneously enables it to uncover correlations and patterns that might otherwise go unnoticed by human researchers.

A notable example of AI's interdisciplinary potential is Microsoft's Project InnerEye, which uses AI-driven imaging tools to assist radiologists in making faster, more accurate diagnoses by enhancing image recognition and diagnostic capabilities. By leveraging historical data patterns, AI systems can forecast emerging trends with remarkable accuracy. Unlike traditional search engines that rely on basic keyword matching, cognitive AI systems interpret context, returning nuanced, meaningful, and highly relevant results— an essential feature for professionals working in R&D-intensive industries. Natural language processing (NLP) capabilities enhance these systems further by transforming unstructured scientific data into actionable insights, summarizing complex research findings, and even generating new hypotheses for researchers to test.

AI's Role in Intellectual Property Protection

The emergence of powerful AI-driven search and information-processing tools is revolutionizing how scientists draft patents and develop intellectual property (IP) strategies. AI now plays a critical role in managing, safeguarding, and expanding the scope of patents and innovations, allowing companies to stay ahead in an increasingly competitive global market.

AI-powered platforms like TurboPatent analyze existing patent landscapes in detail, identifying potential infringements and helping organizations avoid costly legal battles. These systems can quickly assess the novelty of new inventions by comparing proposed innovations against millions of existing patents worldwide, saving time and reducing the risk of unintentional infringement.

Natural language processing (NLP) tools are streamlining the patent application process, reducing the need for extensive legal consultations while accelerating the drafting of comprehensive, well-structured patent applications. This automation not only saves time and resources but also enhances the accuracy and thoroughness of intellectual property documentation.

AI-driven analytics enable companies to devise effective IP strategies by predicting patent expiration trends, monitoring competitors' portfolios, and identifying gaps in the existing IP landscape where new innovations can thrive. IBM's Watson AI, for example, supports businesses by providing powerful analytics tools that help navigate complex IP environments and identify areas ripe for strategic development.

AI for New Material Discovery and Formula Optimization

The discovery of novel materials and the optimization of chemical formulations have historically been labor-intensive processes, often requiring years of trial and error. For many scientists, the journey to becoming proficient in material synthesis and development requires years of specialized education and training. AI is fundamentally changing this dynamic by accelerating every stage of the material design process, from theoretical modeling to experimental validation in the laboratory.

Advanced AI algorithms now have the capacity to predict molecular structures, biological pathways, and physical phenomena before any physical experimentation begins. DeepMind's AlphaFold, for example, has revolutionized the field of protein structure prediction, offering near-instant insights into molecular interactions and significantly accelerating pharmaceutical research and drug discovery.

AI-driven platforms are capable of generating entirely new chemical structures optimized for desired material properties such as mechanical strength, flexibility, thermal stability, or electrical conductivity. Companies like Citrine Informatics leverage machine learning algorithms to design materials tailored to specific functional requirements, dramatically reducing development timelines while enhancing innovation. AI is transforming how formulations are optimized for performance improvements. For example, in the coatings industry, AI-driven insights suggest adjustments to formulations that enhance durability, accelerate drying times, and improve resistance to environmental factors like UV degradation. These advancements lead to the development of eco-friendly products without compromising on quality or performance.

In the near future, formulation development and manufacturing scale-up will increasingly rely on digital twins and material performance simulations conducted in virtual environments. Platforms like NVIDIA's Omniverse enable scientists, engineers, and designers to collaborate on digital twin technology for industrial applications, facilitating real-time performance testing and virtual simulations. This approach dramatically reduces development costs, optimizes resource allocation, and accelerates the pace of innovation.

A landmark study published in *Nature* by the University of Liverpool in November 2024 highlights the potential of AI-powered automation. Researchers demonstrated that AI-driven robots could independently perform complex chemical syntheses, making real-time decisions to alter reaction pathways as needed—achieving outcomes that surpass even those of highly trained PhD chemists. This breakthrough signals a paradigm shift in automated discovery, enabling rapid iteration and unprecedented innovation across chemical research and development.

AI for Manufacturing

The integration of AI into manufacturing processes marks the beginning of the smart factory era, where efficiency, precision, and safety are prioritized at every stage of production. One of the most significant benefits AI brings to modern manufacturing is predictive maintenance. By continuously monitoring equipment in real time, AI systems can anticipate mechanical failures before they occur, minimizing downtime and reducing maintenance costs. Siemens' predictive maintenance technology, for example, has achieved remarkable success, cutting machine downtime in global manufacturing operations by up to 30%.

Real-time AI-driven adjustments to manufacturing processes ensure consistent product quality while minimizing material waste and energy consumption. In semiconductor manufacturing, for instance, AI-enabled robots detect microscopic defects that would otherwise be invisible to the human eye, guaranteeing the structural integrity of each product. Advanced quality control is another transformative advantage. AI-powered vision systems inspect products for defects at scales beyond human capabilities, ensuring compliance with international safety and quality standards.

AI technologies are also enhancing human operator safety by ensuring real-time compliance with workplace safety protocols. These systems track the proper use of personal protective equipment (PPE), monitor hydration levels, and detect signs of fatigue or stress. By proactively identifying potential safety risks, AI minimizes workplace accidents and fosters a safer, more productive working environment.

AI in Supply Chain Optimization

Global supply chains are inherently complex and vulnerable to disruption from geopolitical shifts, economic volatility, and natural disasters. Recent global events have underscored the importance of building more resilient and adaptive supply chains. AI is revolutionizing supply chain management by enhancing both flexibility and predictive accuracy, enabling companies to respond swiftly to dynamic market conditions.

AI-powered demand forecasting models take into account seasonal trends, geopolitical events, and shifts in consumer behavior to deliver highly accurate predictions. For example, Amazon's supply chain network employs sophisticated machine learning algorithms to optimize inventory levels across its vast global distribution centers, reducing both shortages and excess stock while cutting storage costs.

AI-driven route optimization systems adjust delivery routes in real time, considering factors such as traffic patterns, adverse weather conditions, and evolving regulatory constraints. Alibaba's Cainiao logistics network demonstrates the power of AI in improving the efficiency of China's expansive delivery infrastructure, ensuring timely deliveries while reducing carbon emissions.

In highly regulated industries such as pharmaceuticals, chemicals, and food manufacturing, AI systems analyze historical data alongside current legislative trends to predict future regulatory changes. These models help organizations proactively identify safer substitutes for restricted materials, ensuring compliance while minimizing the risk of costly regulatory violations.

AI Governance and Ethical Leadership

As the adoption of AI technologies accelerates, ensuring their ethical use becomes an essential responsibility for technology leaders and policymakers. Ethical leadership in AI demands the creation of transparent frameworks that foster accountability, fairness, and responsible decision-making.

One of the most pressing concerns in AI governance is the issue of explainability. Black-box algorithms, whose internal decision-making processes are opaque, pose significant ethical and legal risks for organizations. These systems must be designed with transparency in mind, allowing users to understand how decisions are made and ensuring accountability at every level.

AI systems trained on biased historical data can unintentionally perpetuate societal inequalities. To address this, organizations must implement rigorous auditing procedures, adopt fairness metrics, and use diverse training datasets to mitigate bias. Clear frameworks for accountability must be established to define who is responsible for the outcomes of AI-driven decisions, ensuring regulatory compliance and fostering public trust.

Data privacy regulations such as the General Data Protection Regulation (GDPR) and the California Consumer Privacy Act (CCPA) require organizations to prioritize user data protection. AI systems should be designed according to privacy-by-design principles, ensuring that data security and transparency remain at the forefront of system architecture.

AI in Innovation and Sustainability: The Final Frontier

AI's role in driving global innovation while advancing sustainability objectives cannot be overstated. As organizations worldwide strive to meet ambitious environmental, social, and governance (ESG) goals, AI offers powerful tools for accelerating the development of sustainable technologies and practices. AI models can predict the environmental impact of new materials and technologies, allowing researchers to develop eco-friendly alternatives proactively. IBM's AI research has led to the discovery of biodegradable polymers capable of replacing conventional plastics, reducing environmental waste and promoting circular economy practices.

AI-driven innovations are revolutionizing recycling processes by improving the accuracy of material separation and enabling more efficient resource reuse. Machine learning algorithms now identify recyclable materials with near-perfect accuracy, dramatically reducing landfill waste and promoting responsible consumption. Platforms like InnoCentive, powered by AI-driven innovation ecosystems, foster collaboration between industries, academic researchers, and startups to address pressing global sustainability challenges, from carbon capture technologies to scalable renewable energy storage solutions.

Conclusion: Embracing the Scientific Revolution

The ongoing scientific revolution driven by AI is not just a movement for technology leaders—it is a call to action for scientists, engineers, researchers, business professionals, and policymakers alike. AI enhances human potential by accelerating discovery cycles, fostering sustainable innovation, and enabling groundbreaking research that addresses global challenges, from climate change mitigation to improved public health outcomes. For business leaders, AI provides powerful tools for data-driven decision-making, streamlining operations, and unlocking new avenues for growth. Beyond its technological implications, AI's transformative impact extends to the human experience. The development of smarter healthcare solutions, sustainable technologies, and personalized customer experiences will significantly improve the quality of life for individuals across the globe.

The future belongs to those who embrace AI as a vital partner in solving humanity's most complex challenges. Whether unlocking the mysteries of the universe or building a more sustainable planet, the true power of AI lies in its ability to amplify human creativity, extend our capabilities, and inspire a collective vision of possibility, progress, and hope.

About the Author

Dr. Polina R. Ware is a visionary leader, connector, and integrator with a passion for aligning technology and business strategies to drive transformative growth. With deep expertise in R&D leadership, AI/

ML applications in chemistry, and digital transformation, she delivers scalable solutions that fuel innovation, efficiency, and sustainability. Polina began her career at 3M, advancing through global leadership roles at LANXESS, Rogers Corporation, and PPG. Her achievements include pioneering novel sustainable coating and adhesive technologies, leading multimillion-dollar M&A integrations, and spearheading AI-driven digital initiatives that optimize performance and accelerate material discovery.

A recognized thought leader, award-winning speaker, and author, Polina champions diversity, inclusion, and high-trust teams that drive industry-changing innovations. She holds a PhD in Polymer Science and Engineering from the University of Massachusetts Amherst and an Executive MBA from Kellogg School of Management. In her free time, Polina enjoys writing, reading leadership books, snowboarding, weightlifting, and creating memorable moments with her son.

Email: Polina.ware@gmail.com

LinkedIn: https://www.linkedin.com/in/polinaware/

TRANSFORMING FINANCIAL SERVICES: LEADING TECHNOLOGY INNOVATION AND OPERATIONAL EXCELLENCE WITH AI

By Svetlana Zavelskaya, MS, MBA
Senior Technology Leader
New York, New York

> *AI is not about replacing humans, it's about amplifying human potential.*
>
> —Ginni Rometty

In this chapter we'll take a look briefly at the history of AI usage in financial services (it might go back in time much further than you expect), at some practical, down-to-earth ways it is utilized, and what you can expect to see in the real world in the nearest future.

Applications of AI in Finance: From Early Applications to Modern Advancements

With the latest advances in AI, people sometimes forget how long AI or some of its aspects existed and have been used in multiple industries, financial services being one of them. The early days of AI in financial services go back to the 1980s to 1990s, where its main focus was around early attempts at fraud detection (basic rule-based systems), credit scoring (early, basic ML models), market risk understanding, and (of course!) firsts attempts in algorithmic trading (those were the initial quantitative strategies). Considering all the benefits technology provides, financial institutions were always looking at it, and investing time and resources to advance AI. It is thought of as an essential competitive advantage tool, whose importance and impact only grows over time, eventually providing competitive advantages to early adaptors.

The main limitations around developing these capabilities were mostly around organizational data silos, the lack of expandability ("black box" models), and limited computational power, requiring tremendous initial investments into physical infrastructure, therefore setting high entrance costs for those interested in the field and requiring significant political power to secure those investments within the organizations of those interested. As data availability increased, computational power grew, AI algorithms matured, and the stage was set for a new wave of innovation.

Evolving AI Landscape: Beyond Chatbots

The first thing that comes to mind nowadays when someone says "AI" is LLMs—large language models—which is what made ChatGPT famous. An LLM is a type of AI that performs language-related tasks by learning patterns from massive amounts of text data, essentially acting like a very powerful language processing tool trained on a huge dataset.

And because language processing is what it does the best, the first and very natural way of using this type of AI is to talk to it. This gave the idea to one of the first (and extremely popular well outside of financial services too) ways to use it—chatbots. Chatbots are LLM-based agents usually additionally trained on proprietary company data and documentation, as this is what they focus on. The next logical step is large written text processing.

As you can imagine, financial services is a highly regulated space, and often those regulations are lengthy, written in a complicated language, and require a lot of time and effort to interpret. So, here is another way for you to use an LLM documentation summary. Imagine instead of, or in addition to, receiving a long cumbersome set of documents from your bank, you get a one-minute written or video summary of what they contain. Wouldn't that be great? And those already exist; be on the lookout in your inbox. These were just two simple examples of LLM use that are already being deployed by many companies of different sizes, helping them to save time and money (you don't need a live representative to pick up a phone call if you can have a chatbot answer customer questions), and optimize processing.

But this is only the beginning. The next step takes AI from simple text processing into large capability data processing. Here are some examples of how it could be used. One of the earlier ideas for using ML/AI was to utilize it for trading. Now imagine taking it a step further: AI-powered robo-advisors providing highly personalized portfolio recommendations and automated rebalancing based on individual risk profiles and financial goals. That's quite a way to democratize financial services!

Another example is AI-powered loan approvals. AI can quickly and efficiently analyze vast amounts of data, including credit history, income, and financial behavior, to make more accurate and inclusive lending decisions. Robotic process automation (RPA) existed for many years, but combined with the power of AI it opens up a new level of possibilities in automating complex, knowledge-intensive tasks, such as document processing, data entry, and customer onboarding.

Risk assessment: sophisticated AI models are employed to assess and manage a wide range of risks, including credit risk, market risk, and operational risk, enabling financial institutions to make more informed and proactive decisions. As dependencies on IT infrastructure and its availability become more and more important, there are many ways companies are employing AI to analyze system logs and performance data to understand and predict failure of IT infrastructure, helping with maintenance and reducing downtime. All the above examples are just slightly touching the surface of AI potential, but I hope you are already imagining a very different future in the field and are as excited about it as I am.

Unlocking Operational Excellence

The day-to-day reality of financial services is not the catchy news about market moves you see on TV, but more day-to-day operations that ensure market stability, fair and accurate reporting, management of multiple components contributing to overall risk, and many other operational tasks that enable money movement, regulatory reporting, as well as other essential activities.

All these mundane, time-consuming, and, let's be honest, often routine and sometimes boring tasks, such as data entry, account reconciliation, and customer onboarding are at the heart of financial services, enabling the smooth operations and reliability of our financial institutions. Automating them will not only increase efficiency and effectiveness of organizations and the people working in them but free those workers up for more strategic and valuable activities, also improving their job satisfaction. Imagine not having to start every morning with the same enormous Excel spreadsheet. On top of that, automating those tasks could also significantly improve accuracy and reduce the risk of mistakes.

As we know, AI operates utilizing a vast array of data, so the quality of that data is extremely important, as it plays a pivotal role in recognizing existing or identifying new market patterns. That data is analyzed to identify anomalies and, therefore, to proactively identify and assess potential risks, either in credit, market, loans, or other areas. At the same time, with data being so crucial in that process, AI can help identify inconsistencies and errors, ensuring its integrity and reliability for decision-making.

As you can imagine, financial service is a complicated and highly regulated business. Every organization has its own set of internal workflows and processes. Many of them were developed decades ago, later updated, changed, and patched to fit new laws and regulations and incorporate sets of new technologies created over those years. All of them create a highly complex communication and data exchange web with a strong emphasis on the security of the data, both stored and while in transit. Add to it the additional opportunity to connect to highly protected internal financial service infrastructures and cross-communication between multiple institutions. AI-powered solutions can suggest options to optimize those workflows and identify potential bottlenecks and overlaps, therefore improving overall process efficiency.

Another important aspect of what AI can bring into the work of finance is its democratization. Currently, the services provided by financial advisors aren't cheap and not available to everybody. Utilizing AI can reduce the costs of an individual's financial management, therefore making it more available and affordable. Providing such personalized services, tailored to customer individual needs and taking into account their personal preferences and goals, would not only improve customer service but also customers loyalty, while increasing profits to the providers of these services.

Another aspect of that democratization is the availability of microloans, small-size loans that people sometimes need, and the market for those is rather limited. With AI performing financial analysis of an individual's historical financial behavior, it can make decisions about issuing a short-term microloan and its amount, which would be a fast, productive, and positive development for society. In addition, AI can also proactively identify potential issues, like new suspicious patterns or anomalies, unusual transactions, etc., enabling the detection and prevention of fraudulent activities in real time.

Navigating the Challenge

We have talked so far about the current and some of the potential benefits AI could bring to the financial services industry. There are many challenges that we have to think about and address, so that the adoption of AI is done responsibly and beneficially to society. Let's take a look at some of them.

We already discuss the importance of the quality of the data used to train AI models, as this is what the model output depends upon. So, let it be just a reminder before we move to the next challenge. As I said, AI model output is based on the data that was used to train the model. So, any kind of flaw or disbalance in the data will lead to an equal disbalance in the model's output. This means that AI will inherit any bias presented in the data it was trained on, leading to potentially unfair or discriminatory outcomes. It is essential to identify and mitigate biases in data and algorithms to ensure fairness and inclusivity.

A lot of regulations have been introduced around AI development and more will come. The regulatory landscape around AI is constantly evolving, with different laws and regulations accepted around the world, often disparate between countries and jurisdictions. Building trust and understanding of AI is crucial for its future. Many AI models, especially those based on deep learning paradigms, are extremely complex and opaque, making it difficult to understand and making their output sometimes unpredictable. Building that trust and understanding requires developing explainable AI (XAI) techniques that can provide insights into the decision-making process of AI models.

AI models are not built in a vacuum, and the data used to train them is often real and current. With so much fraud going on in the industry, ensuring proper security and protection of both AI algorithms and customers data is absolutely crucial for the next phase of AI implementation. It is critical to develop and deploy AI in a responsible and ethical manner, ensuring that AI systems are used for the benefit of society.

Another challenge the industry is facing is the skills gap. While AI itself is not new, the latest developments and rapid adoption of it require new skills and often different thinking, with not everyone being ready to embrace the change. While there are numerous resources to learn AI, not all of them are of the highest quality. Developing an internal talent pool and attracting a skilled workforce is another challenge financial institutions will need to address in order to fully adopt and take advantage of AI capabilities.

The Future of AI in Finance

AI is developing so fast that it is almost impossible now to predict what shape and form it will take in the future, but one thing is clear: AI is here to stay and to transform not only the financial services industry, but many others as well. It is facing challenges, but it is bright, and it is here. Let's talk quickly about some of the emerging trends in AI and how they may impact the industry.

As I mentioned earlier, a lot of concerns are being raised and new regulations are currently being introduced around understanding AI models and what's going on inside of them, so that the output of those models can be understood and explained, and we can understand what's going on inside our creations. XAI techniques are being developed to make AI models more transparent and understandable, building trust and facilitating regulatory compliance.

The integration of AI with modern financial platforms like blockchain can unlock new possibilities for decentralized financial services, also known as DeFi, such as automated trading, risk management, and lending. It presents not only benefits, but risks as well, as it operates outside of traditional centralized financial infrastructures. AI-driven compliance is another step in the AI adoption direction. Here AI is used increasingly to automate compliance processes, such as anti-money laundering (AML) and know-your-customer (KYC) checks, enhancing efficiency and reducing the risk of non-compliance. AI is a great technology, but it doesn't exist in a vacuum. It is connected to other technologies and, more importantly, is utilized by humans.

Human-AI interaction is the next challenge we will need to address. AI is a tool, and one its roles is to help humans and potentially augment human capabilities, not replace them. Human expertise and judgment remain essential in the financial services industry, and AI should be used to enhance, not eliminate, human decision-making.

AI is revolutionizing financial services, demanding responsible innovation with a human focus, transparency and explainability, fairness for all, continuous improvement, and strong ethical governance. I hope this chapter gave you some food for thought and ideas about the current and future state of the technological development of the financial industry and what role AI may play there. I hope you are as excited about it as I am. Feel free to reach out and share your thoughts.

About the Author

Svetlana Zavelskaya is senior technology leader with extensive experience in the financial services and insurance sectors. She has spearheaded service delivery, architecture, engineering, security, and all aspects of operational management.

She is a forward-thinking and results-oriented professional with a strong emphasis on customer service, technological innovation, and operational excellence. Svetlana's unwavering commitment to these principles consistently drives positive business outcomes.

Throughout her career, Svetlana has demonstrated a proven track record of successfully executing and delivering intricate IT projects and strategies. She has been at the forefront of leading technology transformations and organizational change, with a deep understanding of strategic planning and the successful implementation of cutting-edge technologies.

Beyond professional achievements, Svetlana is a dedicated mentor and enthusiastically supports community service initiatives.

Email: svetlana.zavel@gmail.com

LinkedIn: https://www.linkedin.com/in/svetlana-zavelskaya/

APPENDIX A

LIST OF POPULAR AI TOOLS

If you're looking for an AI tool that you can use for various tasks, below are a selection of popular ones. (Note: The creation of this list was assisted by ChatGPT)

ChatGPT (OpenAI)
Use case: Conversational AI, content generation, coding assistance, customer support.
Description: A large language model that can perform tasks like answering questions, generating text, and providing creative writing prompts.
URL: https://chat.openai.com

DALL·E (OpenAI)
Use case: Image generation from text prompts.
Description: An AI model that creates images based on text descriptions, useful in art, marketing, and design.
URL: https://openai.com/dall-e

Microsoft Copilot (Microsoft 365)
Use case: Productivity assistance in Microsoft apps.
Description: An AI tool integrated into Microsoft 365 applications like Word, Excel, PowerPoint, and Outlook to assist with tasks such as drafting, data analysis, and summarizing.
URL: https://www.microsoft.com/en-us/microsoft-365/copilot

Grok (X)
Use case: Conversational AI for Twitter.
Description: An AI chatbot integrated into Twitter (X), designed to assist users in real-time by answering questions, generating tweets, and providing insights, all in a conversational manner.
URL: https://x.com

Gemini (Google)
Use case: Conversational AI and language model.
Description: A multimodal AI model that handles text, images, and other input types. It is expected to enhance search, productivity, and various creative tasks with advanced language understanding.
URL: https://about.google/intl/en/products/gemini/

Claude (Anthropic)
Use case: Conversational AI and assistant tasks.
Description: A language model focused on safe and reliable conversational interactions, used for answering questions, providing insights, and assisting in creative or technical tasks.
URL: https://www.anthropic.com/product

Perplexity
Use case: AI-powered search engine and question-answering.
Description: An AI search engine designed to answer user queries with concise and sourced information, offering both traditional search results and direct answers to questions.
URL: https://www.perplexity.ai

Phind
Use case: AI-powered search engine for developers.
Description: A search engine designed to answer complex technical and programming questions by providing relevant, code-focused answers from documentation, forums, and other resources.
URL: https://www.phind.com

Otter.ai
Use case: Transcription and meeting notes.
Description: A tool that provides real-time transcription of meetings, interviews, and conversations, with AI-generated notes, summaries, and key points.
URL: https://otter.ai

Duolingo (AI-enhanced version)
Use case: Language learning.
Description: A language learning app that uses AI to personalize lessons, predict learner needs, and provide conversational practice with virtual characters.
URL: https://www.duolingo.com

MidJourney
Use case: Artistic and visual content creation.
Description: A tool that generates high-quality art and illustrations from text prompts, used heavily in creative industries.
URL: https://www.midjourney.com

Jasper AI
Use case: Copywriting and marketing content.
Description: An AI-driven platform that helps in generating marketing copy, blog posts, product descriptions, and social media content.
URL: https://www.jasper.ai

Synthesia
Use case: AI-generated video creation.
Description: A tool that lets users create videos with AI avatars and voiceovers, often used for training videos, marketing, and education.
URL: https://www.synthesia.io

Lumen5
Use case: Video content creation.
Description: This AI-powered tool transforms blog posts and articles into shareable videos by automatically matching text with visuals.
URL: https://www.lumen5.com

Grammarly
Use case: Writing assistance and grammar checking.
Description: An AI tool that provides real-time grammar, punctuation, and style suggestions to improve writing quality.
URL: https://www.grammarly.com

Notion AI
Use case: Productivity and organization.
Description: An extension to Notion, it helps with summarizing content, generating ideas, and automating writing processes within a workspace.
URL: https://www.notion.so/product/ai

Pictory
Use case: Video creation.
Description: An AI tool that converts long-form content into short, engaging videos using AI-generated scenes and text overlays.
URL: https://pictory.ai

Hugging Face
Use case: Machine learning model hosting and deployment.
Description: A platform for hosting, sharing, and deploying machine learning models, particularly useful in natural language processing and computer vision.
URL: https://huggingface.co

Runway ML
Use case: Creative AI for video editing, art, and design.
Description: A platform providing creative AI tools for video editing, animation, and design, widely used by artists and filmmakers.
URL: https://runwayml.com

Copy.ai
Use case: Content generation.
Description: An AI-driven platform that generates marketing copy, product descriptions, and other written content for businesses.
URL: https://www.copy.ai

GitHub Copilot (OpenAI)
Use case: Code generation and software development.
Description: An AI-powered code completion tool integrated into code editors like Visual Studio Code, helping developers write code faster by suggesting whole lines or blocks of code.
URL: https://github.com/features/copilot

APPENDIX B

GLOSSARY OF COMMON AI TERMS

Many of the terms below are beyond the scope of this book. For someone new to AI, just reading through these terms will greatly expand your vision of the workings of AI. This glossary can also provide a starting point for anyone wanting to look for further information about AI. (Note: The creation of this glossary was assisted by ChatGPT)

Activation Function
A mathematical function used in neural networks to determine the output of a node (neuron). Activation functions introduce non-linearity, allowing neural networks to solve more complex problems.

AI-Assisted Creativity
The use of AI tools to enhance human creativity. This includes generating art, music, writing, or design based on user input, making the creative process faster and more accessible.

AI Compass
A strategic framework that guides individuals or organizations in how to ethically and effectively use AI technologies. It helps align AI development with long-term goals, ethical standards, and societal values.

AI-Driven Insights
The ability of AI to analyze large datasets and extract meaningful patterns or trends. These insights help businesses and individuals make data-driven decisions faster and more accurately.

AI GamePlan
A structured plan or strategy that outlines how AI will be developed, integrated, and utilized in specific projects or across an organization. An AI GamePlan typically includes objectives, resources, timelines, and risk management.

AI Governance
The policies, frameworks, and structures that ensure the responsible development, deployment, and management of AI systems. AI governance includes legal, ethical, and operational standards to minimize risks and ensure fairness and transparency.

AI Integration
The process of embedding AI technologies into existing business processes, tools, or workflows. AI integration aims to improve efficiency, accuracy, and decision-making across various industries.

AI-Powered Automation
Using AI to automate repetitive or time-consuming tasks. This can range from chatbots responding to customer inquiries to AI systems automating complex business processes, saving time and resources.

AI Transparency
The practice of making AI systems and their decision-making processes understandable to users. Transparency helps build trust in AI by allowing users to see how and why a model arrived at a certain conclusion.

Algorithm
A set of rules or instructions that a computer follows to perform a task or solve a problem. In AI, algorithms are used to process data, learn from patterns, and make decisions.

Artificial General Intelligence (AGI)
A type of AI that can understand, learn, and apply knowledge across a wide range of tasks at a human-like level. Unlike narrow AI, which excels at specific tasks, AGI would have general cognitive abilities akin to human intelligence.

Augmented Intelligence
An approach where AI is designed to assist and enhance human decision-making rather than replace it. Augmented intelligence focuses on collaboration between humans and AI systems for better outcomes.

Autonomous AI Agents
AI-powered systems that can make independent decisions and take actions without human intervention. Think of AI personal assistants that can plan your schedule, book travel, or even manage entire workflows autonomously.

Backpropagation
An algorithm used to train neural networks by adjusting the weights of connections between neurons. Backpropagation helps minimize errors by propagating them backward through the network.

Bias in AI
Refers to the presence of prejudiced or unfair outcomes in AI models due to biased training data or flawed algorithms. Bias in AI can lead to discrimination or unfair treatment in areas like hiring, lending, and law enforcement.

Convolutional Neural Networks (CNNs)
A type of neural network particularly effective for image and video analysis. CNNs use convolutional layers to capture spatial hierarchies in data, making them well-suited for tasks like object detection and image classification.

Conversational AI
AI systems designed to engage in human-like dialogue. These systems are often used in chatbots, virtual assistants, and customer service applications to respond to user queries and hold natural conversations.

Deep Learning
A type of machine learning that uses neural networks with many layers (hence "deep"). Deep learning excels in processing large, complex datasets and is used in tasks like image and speech recognition, natural language processing, and more.

Edge AI
AI that processes data on local devices (e.g., smartphones, cameras) rather than in the cloud. Edge AI enables faster processing and reduces the need for constant internet connectivity, making it suitable for real-time applications.

Epoch
A term used in machine learning to describe one complete cycle through the entire training dataset. Multiple epochs are used during training to improve the model's accuracy.

Ethics in AI
The principles and guidelines that govern the responsible development and use of AI technologies. Ethical AI focuses on fairness, transparency, accountability, and ensuring AI does no harm to individuals or society.

Explainable AI (XAI)
Techniques and tools that allow humans to understand how AI models make decisions. XAI is important for increasing transparency and trust, particularly in critical applications like healthcare and finance.

Fine-Tuning
A process in machine learning where a pre-trained model is further trained on a smaller, task-specific dataset. Fine-tuning helps adapt a general-purpose model to a specialized application.

Generative AI
AI models designed to generate new content such as text, images, audio, or video. These models can create outputs that resemble human-made content by learning from vast datasets.

Gradient Descent
An optimization algorithm used to minimize errors in machine learning models by adjusting model parameters (like weights in neural networks). It is essential for training models by finding the best possible performance on a task.

Human-in-the-Loop (HITL)
A process where human judgment is integrated with AI decision-making to improve performance. HITL ensures that human oversight guides AI systems in tasks requiring nuanced understanding or critical outcomes.

Hyperparameters
Settings or configurations that define the structure and learning process of a machine learning model (e.g., learning rate, number of layers). Hyperparameters are tuned to optimize model performance.

Inference
The process of applying a trained AI model to new data to make predictions or decisions. Inference is what happens when an AI model is used in real-world applications after training.

Large Language Models (LLMs)
AI models, like GPT-4, that process and generate human-like text. They are trained on massive datasets of text to understand and predict language patterns, enabling a wide range of applications like answering questions, translation, and content generation.

Loss Function
A mathematical function that quantifies the difference between a model's predictions and the actual outcomes. The goal of training is to minimize the loss function, improving the model's accuracy.

Machine Learning (ML)
A subset of AI where algorithms improve their performance at tasks by learning from data, without being explicitly programmed for each specific task. It involves training models to recognize patterns and make decisions based on data.

Natural Language Processing (NLP)
A field of AI focused on enabling computers to understand, interpret, and generate human language. It encompasses tasks like translation, sentiment analysis, summarization, and conversation generation.

Natural Language Understanding (NLU)
A subfield of NLP focused on enabling machines to comprehend the meaning and intent behind human language. NLU is critical for tasks like question answering, sentiment analysis, and dialogue systems.

Neural Networks
Computational models inspired by the human brain, composed of layers of interconnected nodes ("neurons"). Neural networks are the foundation of most AI models and are used to recognize patterns, classify data, and make decisions.

Neural Radiance Fields (NeRFs)
A groundbreaking approach to generating 3D models from simple 2D images, used in virtual reality, digital twins, and high-quality 3D rendering.

Nirvana State
A theoretical concept in AI where a system reaches a level of perfection in its performance, free of bias, errors, and inefficiencies. Achieving this state would mean that the AI operates at optimal levels in all aspects, but it is largely seen as an unattainable ideal.

Overfitting
A situation where a machine learning model learns the training data too well, including its noise and errors. Overfitting causes poor performance on new, unseen data, as the model fails to generalize.

Personalization
The use of AI to tailor experiences to individual users, often by analyzing user data and preferences. Personalization is common in recommendation systems (e.g., Netflix, Spotify) and digital marketing.

Prioritization Matrix
A tool used in decision-making to rank AI projects or tasks based on factors such as urgency, impact, and resources. This matrix helps teams focus on high-priority initiatives and allocate resources effectively.

Prompt Engineering
The practice of designing and refining inputs (prompts) for AI models, particularly large language models, to elicit the desired output. It's key to making generative AI models perform specific tasks effectively.

Quantum Machine Learning
The integration of AI with quantum computing has the potential to solve complex problems far beyond the capabilities of classical computing, with breakthroughs in materials science and cryptography.

Recurrent Neural Networks (RNNs)
A neural network model designed to process sequential data, such as time series or natural language. RNNs have a memory component that helps them maintain context over sequences, often used in speech recognition and language modeling.

Reinforcement Learning
An area of machine learning where an agent learns to make decisions by interacting with its environment and receiving rewards or penalties based on its actions. This method is often used in gaming, robotics, and complex decision-making scenarios.

Reinforcement Learning from Human Feedback (RLHF)
A type of reinforcement learning where AI models are trained and fine-tuned using human input on their outputs. Humans provide feedback, which the model uses to adjust its actions and improve over time, particularly for tasks requiring nuanced judgment.

Supervised Learning
A machine learning approach where a model is trained on labeled data. The model learns to map inputs to specific outputs by observing examples (e.g., classifying images of cats and dogs).

Training Data
The dataset used to teach an AI model. During training, the model learns from this data to make predictions or decisions. The quality and quantity of training data significantly impact the model's performance.

Transfer Learning

A machine learning technique where a model trained on one task is adapted to perform a different but related task. This allows for faster training and improved performance by leveraging previously learned knowledge.

Transformers

A neural network architecture designed to handle sequential data but more efficiently than RNNs. Transformers are the foundation of many modern AI models, including large language models, by enabling parallel processing of data sequences.

Underfitting

Occurs when a machine learning model is too simple to capture the underlying patterns in the data. An underfitted model performs poorly on both the training data and new data.

Unsupervised Learning

A machine learning technique where models are trained on unlabeled data. The goal is to find hidden patterns or groupings in the data without explicit instructions (e.g., clustering similar items).

DID YOU ENJOY THIS BOOK?

If you enjoyed reading this book, you can help by suggesting it to someone else you think might like it, and **please leave a positive review** wherever you purchased it. This does a lot in helping others find the book. We thank you in advance for taking a few moments to do this.

THANK YOU

You might also like other Thin Leaf Press titles:

The AI Mindset: Thriving Within Civilization's Next Big Disruption

AI: Work Smarter and Live Better Within Civilization's Next Big Disruption

Peak Performance: Mindset Tools for Managers

Peak Performance: Mindset Tools for Sales

Peak Performance: Mindset Tools for Leaders

Peak Performance: Mindset Tools for Business

Peak Performance: Mindset Tools for Entrepreneurs

Peak Performance: Mindset Tools for Athletes

The Successful Mind: Tools to Living a Purposeful, Productive, and Happy Life

The Successful Body: Using Fitness, Nutrition, and Mindset to Live Better

The Successful Spirit: Top Performers Share Secrets to a Winning Mindset

Winning Mindset: Elite Strategies for Peak Performance

Winner's Mindset: Peak Performance Strategies for Success

The Life Coach's Tool Kit, Vol. 1

The Life Coach's Tool Kit, Vol. 2

The Life Coach's Tool Kit, Vol. 3

Ordinary to Extraordinary

The Magical Lightness of Being

Explore.

www.ingramcontent.com/pod-product-compliance
Lightning Source LLC
Chambersburg PA
CBHW071315210326
41597CB00015B/1235